BIRKHÄUSER

Patrick R. Girard

Quaternions, Clifford Algebras and Relativistic Physics

Birkhäuser
Basel · Boston · Berlin

Author:

Patrick R. Girard
INSA de Lyon
Département Premier Cycle
20, avenue Albert Einstein
F-69621 Villeurbanne Cedex
France
e-mail: patrick.girard@insa-lyon.fr

Igor Ya. Subbotin
Department of Mathematics and Natural Sciences
National University
Los Angeles Campus
5245 Pacific Concourse Drive
Los Angeles, CA 90045
USA
e-mail: isubboti@nu.edu

2000 Mathematical Subject Classification: 15A66, 20G20, 30G35, 35Q75, 78A25, 83A05, 83C05, 83C10

Library of Congress Control Number: 2006939566

Bibliographic information published by Die Deutsche Bibliothek
Die Deutsche Bibliothek lists this publication in the Deutsche Nationalbibliografie; detailed bibliographic data is available in the Internet at <http://dnb.ddb.de>.

ISBN 978-3-7643-7790-8 Birkhäuser Verlag AG, Basel – Boston – Berlin

Part of Springer Science+Business Media
Printed on acid-free paper produced from chlorine-free pulp. TCF ∞
Printed in Germany
ISBN-10: 3-7643-7790-8
ISBN-13: 978-3-7643-7790-8

e-ISBN-10: 3-7643-7791-7
e-ISBN-13: 978-3-7643-7791-5

9 8 7 6 5 4 3 2 1

www.birkhauser.ch

To Isabelle, my wife, and to our children: Claire, Béatrice, Thomas and Benoît

Foreword

The use of Clifford algebra in mathematical physics and engineering has grown rapidly in recent years. Clifford had shown in 1878 the equivalence of two approaches to Clifford algebras: a geometrical one based on the work of Grassmann and an algebraic one using tensor products of quaternion algebras $\mathbb{H}$. Recent developments have favored the geometric approach (*geometric algebra*) leading to an algebra (*space-time algebra*) complexified by the algebra $\mathbb{H} \otimes \mathbb{H}$ presented below and thus distinct from it. The book proposes to use the algebraic approach and to define the Clifford algebra intrinsically, independently of any particular matrix representation, as a tensor product of quaternion algebras or as a subalgebra of such a product. The quaternion group thus appears as a fundamental structure of physics.

One of the main objectives of the book is to provide a pedagogical introduction to this new calculus, starting from the quaternion group, with applications to physics. The volume is intended for professors, researchers and students in physics and engineering, interested in the use of this new quaternionic Clifford calculus.

The book presents the main concepts in the domain of, in particular, the quaternion algebra $\mathbb{H}$, complex quaternions $\mathbb{H}(\mathbb{C})$, the Clifford algebra $\mathbb{H} \otimes \mathbb{H}$ real and complex, the multivector calculus and the symmetry groups: $SO(3)$, the Lorentz group, the unitary group $SU(4)$ and the symplectic unitary group $USp(2, \mathbb{H})$. Among the applications in physics, we examine in particular, special relativity, classical electromagnetism and general relativity.

I want to thank G. Casanova for having confirmed the validity of the interior and exterior products used in this book, F. Sommen for a confirmation of the Clifford theorem and A. Solomon for having attracted my attention, many years ago, to the quaternion formulation of the symplectic unitary group.

Further thanks go to Professor Bernard Balland for reading the manuscript, the Docinsa library, the computer center and my colleagues: M.-P. Noutary for advice concerning *Mathematica*, G. Travin and A. Valentin for their help in Latex.

For having initiated the project of this book in a conversation, I want to thank the Presses Polytechniques et Universitaires Romandes, in particular, P.-F. Pittet and O. Babel.

Finally, for the publication of the english translation, I want to thank Thomas Hempfling at Birkhäuser.

Lyon, June 2006

Patrick R. Girard

Contents

Introduction

If one examines the mathematical tools used in physics, one finds essentially three calculi: the classical vector calculus, the tensor calculus and the spinor calculus. The three-dimensional vector calculus is used in nonrelativistic physics and also in classical electromagnetism which is a relativistic theory. This calculus, however, cannot describe the unity of the electromagnetic field and its relativistic features. As an example, a phenomenon as simple as the creation of a magnetic induction by a wire with a current is in fact a purely relativistic effect. A satisfactory treatment of classical electromagnetism, special relativity and general relativity is given by the tensor calculus. Yet, the tensor calculus does not allow a double representation of the Lorentz group and thus seems incompatible with relativistic quantum mechanics. A third calculus is then introduced, the spinor calculus, to formulate relativistic quantum mechanics. The set of mathematical tools used in physics thus appears as a succession of more or less coherent formalisms. Is it possible to introduce more coherence and unity in this set? The answer seems to reside in the use of Clifford algebra. One of the major benefits of Clifford algebras is that they yield a simple representation of the main covariance groups of physics: the rotation group $SO(3)$, the Lorentz group, the unitary and symplectic unitary groups. Concerning $SO(3)$, this is well known, since the quaternion algebra $\mathbb{H}$ which is a Clifford algebra (with two generators) allows an excellent representation of that group . The Clifford algebra $\mathbb{H} \otimes \mathbb{H}$, the elements of which are simply quaternions having quaternions as coefficients, allows us to do the same for the Lorentz group. One shall notice that $\mathbb{H} \otimes \mathbb{H}$ is defined intrinsically independently of any particular matrix representation. By taking $\mathbb{H} \otimes \mathbb{H}$ (over $\mathbb{C}$), one obtains the Dirac algebra and a simple representation of $SU(4)$ and $USp(2, \mathbb{H})$. Computations within these algebras have become straightforward with software like *Mathematica* which allows us to perform extended algebraic computations and to simplify them. One will find as appendices, worksheets which allow easy programming of the algebraic (or numerical) calculi presented here. One of the main objectives of this book is to show the interest in the use of Clifford algebra $\mathbb{H} \otimes \mathbb{H}$ in relativistic physics with applications such as classical electromagnetism, special relativity and general relativity.

Chapter 1

Quaternions

The abstract quaternion group, discovered by William Rowan Hamilton in 1843, is an illustration of group structure. After having defined this fundamental concept of physics, the chapter examines as examples the finite groups of order $n \leq 8$ and in particular, the quaternion group. Then the quaternion algebra and the classical vector calculus are treated as an application.

1.1 Group structure

A set G of elements is a group if there exists an internal composition law $*$ defined for all elements and satisfying the following properties:

1. the law is associative

$$(a * b) * c = a * (b * c), \qquad \forall a, b, c \in G,$$

2. the law admits an identity element e

$$a * e = e * a = a, \qquad \forall a \in G,$$

3. any element a of G has an inverse a'

$$a' * a = a * a' = e.$$

Let F and G be two groups. A composition law on $F \times G$ is defined by

$$(f_1, g_1)(f_2, g_2) = (f_1 f_2, g_1 g_2), \qquad (f_i \in F,\ g_i \in G,\ i = 1, 2);$$

the group $F \times G$ is called the direct product of the groups F and G.

Examples. 1. Cyclic group C_n of order n the elements of which are

$$(b, b^2, b^3, \ldots, b^n = e)$$

and where b represents, for example, a rotation of $2\pi/n$ around an axis.

2. Dihedral group D_n of order $2n$ generated by two elements a and b such that

$$a^2 = b^n = (ab)^2 = e.$$

One has in particular $b^{-h}a = ab^h$ $(h = 1 \cdots n)$; indeed, since

$$(ab)^{-1} = b^{-1}a^{-1} = b^{-1}a = ab,$$

one has

$$b^{-1}(b^{-1}a)b = b^{-2}ab = b^{-1}ab^2 = ab^3$$

and thus $b^{-2}a = ab^2$; by proceeding similarly by recurrence, one establishes the above formula.

1.2 Finite groups of order $n \leq 8$

The finite groups of order $n \leq 8$, except the quaternion group which will be treated separately, are the following.

1. $n = 1$, there exists only one group $(1 = e)$.
2. $n = 2$, only one group exists, the cyclic group C_2 consisting of the elements $(b,\, b^2 = e)$.

   Examples. (a) the group constituted by the elements $(-1,\, 1)$;

   (b) the group having the elements (b : rotation of $\pm\pi$ around an axis, $b^2 = e$).

3. $n = 3$, only one group is possible: the cyclic group C_3 of elements $(b,\, b^2,\, b^3 = e)$ where b, b^2 are elements of order 3.
4. $n = 4$, two groups exist:

   (a) the cyclic group C_4 constituted by the elements $(b,\, b^2,\, b^3,\, b^4 = e)$ where the element b^2 is of order 2, and where $(b,\, b^3)$ are elements of order 4;

   (b) the Klein four-group defined by

   $$I^2 = J^2 = (IJ)^2 = 1$$

   or, equivalently $I^2 = J^2 = K^2 = IJK = 1$ with $K = IJ$ and the multiplication table

	1	I	J	K
1	1	I	J	K
I	I	1	K	J
J	J	K	1	I
K	K	J	I	1

.

The Klein four-group is isomorphic to the direct product of two cyclic groups C_2,

$$(-1,1) \times (b, b^2 = e)$$
$$= \{1 \equiv (1, b^2),\ \ I \equiv (1, b),\ \ J \equiv (-1, b),\ \ K \equiv (-1, b^2)\}\,.$$

Example. The group constituted by the elements (I: rotation of π around the axis Ox, J: rotation of π around the axis Oy, $K = IJ$: rotation of π around the axis Oz).

5. $n = 5$, there exists only one group, the cyclic group C_5 having the elements (b, b^2, b^3, b^4, $b^5 = e$).

6. $n = 6$, two groups are possible:

   (a) the cyclic group C_6 (b, b^2, b^3, b^4, b^5, $b^6 = e$);

   (b) the dihedral group D_3 defined by the relations

$$a^2 = b^3 = (ab)^2 = e,$$

leading to the multiplication table

	b	b^2	$b^3 = e$	a	ab	ba
b	b^2	e	b	ba	a	ab
b^2	e	b	b^2	ab	ba	a
$b^3 = e$	b	b^2	e	a	ab	ba
a	ab	ba	a	e	b	b^2
ab	ba	a	ab	b^2	e	b
ba	b	ab	ba	b	b^2	e

with $b^{-h}a = ab^h$ ($h = 1, 2, 3$). This group is the first noncommutative group of the series.
Example. The symmetry group of the equilateral triangle (see Fig. 1.1).

7. $n = 7$, there exists only one group, the cyclic group C_7 of elements (b, b^2, b^3, b^4, b^5, b^6, $b^7 = e$).

8. $n = 8$, there exist five groups, among them the quaternion group which will be treated separately.

   (a) The cyclic group C_8 of elements (b, b^2, b^3, b^4, b^5, b^6, b^7, $b^8 = e$).

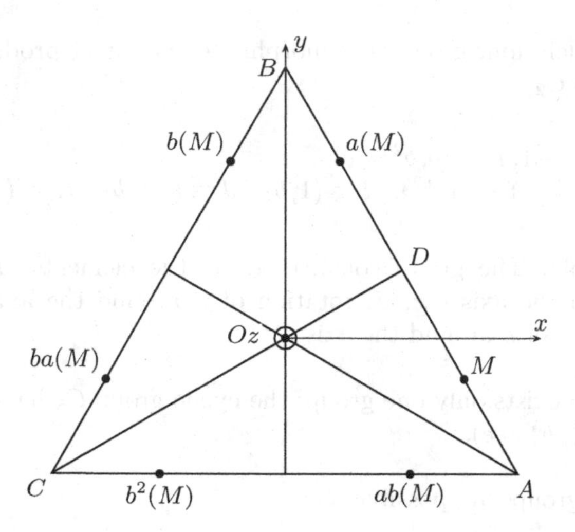

Figure 1.1: Symmetry group of the equilateral triangle; b represents a rotation of the point M by $2\pi/3$ around the axis Oz, a a symmetry of M in the plane ABC with respect to the mediatrice CD and M an arbitrary point of the triangle.

(b) The group $S_{2\times2\times2}$, direct product of the Klein four-group with C_2,

$$(1, I, J, K) \times (1, -1) = (\pm 1, \pm I, \pm J, \pm K)$$

with $1 = (1,1)$, $-1 = (1,-1)$, $\pm I = (I, \pm 1)$, $\pm J = (J, \pm 1)$, $\pm K = (K, \pm 1)$; the multiplication table is given by

	1	-1	I	$-I$	J	$-J$	K	$-K$
1	1	-1	I	$-I$	J	$-J$	K	$-K$
-1	-1	1	$-I$	I	$-J$	J	$-K$	K
I	I	$-I$	1	-1	K	$-K$	J	$-J$
$-I$	$-I$	I	-1	1	$-K$	K	$-J$	J
J	J	$-J$	K	$-K$	1	-1	I	$-I$
$-J$	$-J$	J	$-K$	K	-1	1	$-I$	I
K	K	$-K$	J	$-J$	I	$-I$	1	-1
$-K$	$-K$	K	$-J$	J	$-I$	I	-1	1

;

the group is commutative.

(c) the group $S_{4\times2}$, direct product of C_4 with C_2 and constituted by the elements

$$(b, b^2, b^3, b^4 = e) \times (1, -1) = (\pm b, \pm b^2, \pm b^3, \pm 1);$$

it is a commutative group.

(d) The group D_4 (noncommutative) defined by

$$a^2 = b^4 = (ab)^2 = e$$

with the multiplication table

	b	b^2	b^3	$b^4 = e$	a	ab	ba	ab^2
b	b^2	b^3	e	b	ba	a	ab^2	ab
b^2	b^3	e	b	b^2	ab^2	ba	ab	a
b^3	e	b	b^2	b^3	ab	ab^2	a	ba
$b^4 = e$	b	b^2	b^3	e	a	ab	ba	ab^2
a	ab	ab^2	ba	a	e	b	b^3	b^2
ab	ab^2	ba	a	ab	b^3	e	b^2	b
ba	a	ab	b^2	ba	b	b^2	e	b^3
ab^2	ab	a	ab	ab^2	b^2	b^3	b	e

and $b^{-h}a = ab^h$ $(h = 1, 2, 3, 4)$.
Example. The symmetry group of the square (see Fig. 1.2).

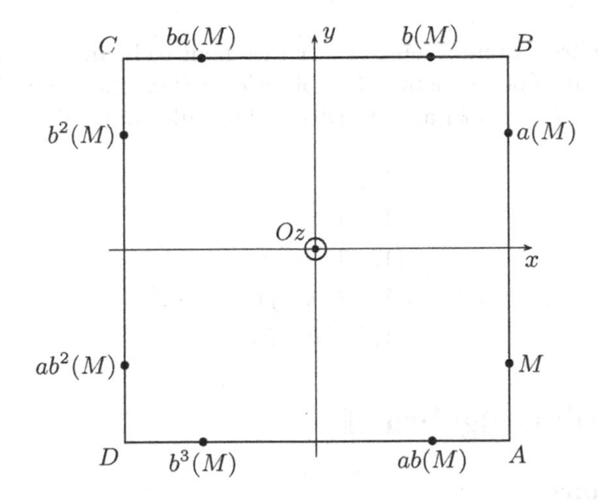

Figure 1.2: Symmetry group of the square; b is a rotation of $\pi/2$ around the axis Oz, a a symmetry with respect to the axis Ox in the plane $ABCD$ and M an arbitrary point of the square.

1.3 Quaternion group

The quaternion group (denoted Q) was discovered by William Rowan Hamilton in 1843 and is constituted by the eight elements ± 1, $\pm i$, $\pm j$, $\pm k$ satisfying the relations

$$
\begin{aligned}
i^2 = j^2 = k^2 = ijk = -1,\\
ij = -ji = k,\\
jk = -kj = i,\\
ki = -ik = j,
\end{aligned}
$$

with the multiplication table

	1	-1	i	$-i$	j	$-j$	k	$-k$
1	1	-1	i	$-i$	j	$-j$	k	$-k$
-1	-1	1	$-i$	i	$-j$	j	$-k$	k
i	i	$-i$	-1	1	k	$-k$	$-j$	j
$-i$	$-i$	i	1	-1	$-k$	k	j	$-j$
j	j	$-j$	$-k$	k	-1	1	i	$-i$
$-j$	$-j$	j	k	$-k$	1	-1	$-i$	i
k	k	$-k$	j	$-j$	$-i$	i	-1	1
$-k$	$-k$	k	$-j$	j	i	$-i$	1	-1

the element of the first column being the first element to be multiplied and 1 being the identity element. The element -1 is of order 2 (i.e., its square is equal to 1) and the elements $(\pm i, \pm j, \pm k)$ are of order 4. The subgroups of Q are

$$
\begin{aligned}
&(1)\\
&(1, -1)\\
&(1, -1, i, -i)\\
&(1, -1, j, -j)\\
&(1, -1, k, -k).
\end{aligned}
$$

1.4 Quaternion algebra $\mathbb{H}$

1.4.1 Definitions

Consider the vector space of numbers called quaternions $a, b, \ldots$ constituted by four real numbers

$$
\begin{aligned}
a &= a_0 + a_1 i + a_2 j + a_3 k\\
&= (a_0, a_1, a_2, a_3)\\
&= (a_0, \mathbf{a}) = (a_0, \underline{a})
\end{aligned}
$$

where $S(a) = a_0$ is the scalar part and $V(a) = \mathbf{a} = \underline{a}$ the vectorial part. This vector space is transformed into the associative algebra of quaternions (denoted $\mathbb{H}$) via the multiplication

$$\begin{aligned} ab = & (a_0b_0 - a_1b_1 - a_2b_2 - a_3b_3) \\ & + (a_0b_1 + a_1b_0 + a_2b_3 - a_3b_2)i \\ & + (a_0b_2 + a_2b_0 + a_3b_1 - a_1b_3)j \\ & + (a_0b_3 + a_3b_0 + a_1b_2 - a_2b_1)k \end{aligned}$$

and in a more condensed form

$$ab = (a_0b_0 - \mathbf{a} \cdot \mathbf{b}, a_0\mathbf{b} + b_0\mathbf{a} + \mathbf{a} \times \mathbf{b})$$

where $\mathbf{a} \cdot \mathbf{b} = (a_1b_1 + a_2b_2 + a_3b_3)$ and $\mathbf{a} \times \mathbf{b} = (a_2b_3 - a_3b_2)i + (a_3b_1 - a_1b_3)j + (a_1b_2 - a_2b_1)k$ are respectively the usual scalar and vector products. Historically, these two products were obtained by W. J. Gibbs [17] by taking $a_0 = b_0 = 0$ and by separating the quaternion product in two parts.

The quaternion algebra constitutes a noncommutative field (without divisors of zero) containing $\mathbb{R}$ and $\mathbb{C}$ as particular cases. Let $a = a_0 + a_1i + a_2j + a_3k$ be a quaternion, the conjugate of a, the square of its norm and its norm are respectively

$$a_c = a_0 - a_1i - a_2j - a_3k,$$

$$|a|^2 = aa_c = a_0^2 + a_1^2 + a_2^2 + a_3^2,$$

$$|a| = \sqrt{|a|^2},$$

with the following properties

$$(ab)_c = b_ca_c,$$

$$|ab|^2 = |a|^2 |b|^2 ,$$

the last relation deriving from $(ab)_cab = b_ca_cab = (aa_c)(bb_c)$; furthermore,

$$S(ab) = S(ba)$$

thus $S[a(bc)] = S[(bc)a] = S[b(ca)] = S[(ca)b]$ and therefore

$$S(abc) = S(bca) = S(cab),$$

$$a^{-1} = \frac{a_c}{aa_c},$$

$$|a^{-1}|^2 = \frac{a_c}{aa_c}\left(\frac{a}{aa_c}\right) = \frac{1}{|a|^2},$$

$$(a_1a_2\cdots a_n)^{-1} = \frac{(a_1a_2\cdots a_n)_c}{|a_1a_2\cdots a_n|^2} = a_n^{-1}a_{n-1}^{-1}\cdots a_1^{-1}.$$

To divide a quaternion a by the quaternion b ($\neq 0$), one simply has to resolve the equation

$$xb = a \qquad \text{or} \qquad by = a$$

with the respective solutions

$$x = ab^{-1} = a\frac{b_c}{|b|^2},$$

$$y = b^{-1}a = \frac{b_c a}{|b|^2}$$

and the relation $|x| = |y| = \frac{|a|}{|b|}$.

Examples. Consider the quaternions $a = 2 + 4i - 3j + k$ and $b = 5 - 2i + j - 3k$;

1. the vectorial parts $\underline{a}$, $\underline{b}$ and the conjugates a_c, b_c are

$$\underline{a} = 4i - 3j + k, \qquad \underline{b} = -2i + j - 3k,$$
$$a_c = 2 - 4i + 3j - k, \qquad b_c = 5 + 2i - j + 3k;$$

2. the norms are given by

$$|a| = \sqrt{aa_c} = \sqrt{30},$$
$$|b| = \sqrt{bb_c} = \sqrt{39};$$

3. the inverses are

$$a^{-1} = \frac{a_c}{|a|^2} = \frac{2 - 4i + 3j - k}{30},$$
$$b^{-1} = \frac{b_c}{|b|^2} = \frac{5 + 2i - j + 3k}{39};$$

4. one can realize the following operations

$$a + b = 7 + 2i - 2j - 2k,$$
$$a - b = -3 + 6i - 4j + 4k,$$
$$ab = 24 + 24i - 3j - 3k, \qquad |ab| = \sqrt{1\,170},$$
$$ba = 24 + 8i - 23j + k, \qquad |ba| = \sqrt{1\,170},$$
$$(ab)^{-1} = \frac{(ab)_c}{|ab|^2} = b^{-1}a^{-1} = \frac{4}{195} - \frac{4i}{195} + \frac{j}{390} + \frac{k}{390},$$

$$xa = b, \qquad x = ba^{-1} = (\frac{-2}{15} - \frac{8i}{15} + \frac{9j}{10} - \frac{13k}{30}),$$
$$ay = b, \qquad y = a^{-1}b = (\frac{-2}{15} - \frac{16i}{15} + \frac{7j}{30} - \frac{3k}{10}),$$
$$S(x) = S(y), \qquad |x| = |y| = \sqrt{\frac{13}{10}}.$$

1.4.2 Polar form

Any nonzero quaternion can be written

$$\begin{aligned} a &= a_0 + a_1 i + a_2 j + a_3 k \\ &= r(\cos\theta + \mathbf{u}\sin\theta), \qquad 0 \le \theta \le 2\pi \end{aligned}$$

with $r = |a| = \sqrt{a_0^2 + a_1^2 + a_2^2 + a_3^2}$ being the norm of a and

$$\cos\theta = \frac{a_0}{r}, \qquad \sin\theta = \frac{\pm\sqrt{a_1^2 + a_2^2 + a_3^2}}{r},$$
$$\cot\theta = \pm\frac{a_0}{|\underline{a}|}, \qquad \tan\theta = \pm\frac{|\underline{a}|}{a_0};$$

the unit vector $\mathbf{u}$ ($\mathbf{u}\mathbf{u}_c = 1$) is given by

$$\mathbf{u} = \frac{\pm(a_1 i + a_2 j + a_3 k)}{\sqrt{a_1^2 + a_2^2 + a_3^2}}$$

with $a_1^2 + a_2^2 + a_3^2 \neq 0$. Since $\mathbf{u}^2 = -1$, one has via the De Moivre theorem

$$a^n = r^n(\cos n\theta + \mathbf{u}\sin n\theta).$$

Example. Consider the quaternion a; let us determine its polar form

$$\begin{aligned} a &= 3 + i + j + k, \\ |a| &= \sqrt{12} = 2\sqrt{3}, \qquad |\underline{a}| = \sqrt{3}, \\ \tan\theta &= \frac{|\underline{a}|}{a_0} = \frac{1}{\sqrt{3}}, \qquad \theta = 30^\circ, \\ \text{Answer: } a &= 2\sqrt{3}\left[\cos 30^\circ + \left(\frac{i+j+k}{\sqrt{3}}\right)\sin 30^\circ\right]. \end{aligned}$$

1.4.3 Square root and n^{th} root

Square root

The square root of a quaternion $a = a_0 + a_1 i + a_2 j + a_3 k$ can be obtained algebraically as follows. The equation $b^2 = a$ with $b = b_0 + b_1 i + b_2 j + b_3 k$ leads to the following equations

$$b_0^2 - b_1^2 - b_2^2 - b_3^2 = a_0 \tag{1.1}$$
$$2b_0 b_1 = a_1, \qquad \operatorname{sgn}(b_0 b_1) = \operatorname{sgn}(a_1), \tag{1.2}$$
$$2b_0 b_2 = a_2, \qquad \operatorname{sgn}(b_0 b_2) = \operatorname{sgn}(a_2), \tag{1.3}$$
$$2b_0 b_3 = a_3, \qquad \operatorname{sgn}(b_0 b_3) = \operatorname{sgn}(a_3). \tag{1.4}$$

Writing $t = b_0^2$ the above equation (1.1) leads to

$$t - \frac{a_1^2 + a_2^2 + a_3^2}{4t} = a_0$$

or

$$t^2 - a_0 t - \frac{a_1^2 + a_2^2 + a_3^2}{4} = 0.$$

One obtains

$$t = b_0^2 = \frac{a_0 + \sqrt{a_0^2 + a_1^2 + a_2^2 + a_3^2}}{2} \geq 0$$
$$-(b_1^2 + b_2^2 + b_3^2) = a_0 - b_0^2 = \frac{a_0 - \sqrt{a_0^2 + a_1^2 + a_2^2 + a_3^2}}{2},$$

hence

$$b_0 = \frac{\varepsilon}{\sqrt{2}} \sqrt{\sqrt{a_0^2 + a_1^2 + a_2^2 + a_3^2} + a_0} \qquad (\varepsilon = \pm 1).$$

The equations (1.2), (1.3), (1.4) lead to

$$b_0^2(-b_1^2) = \frac{-a_1^2}{4}, \tag{1.5}$$
$$b_0^2(-b_2^2 - b_3^2) = \frac{-(a_2^2 + a_3^2)}{4}, \tag{1.6}$$

with

$$\begin{aligned} -b_2^2 - b_3^2 &= b_1^2 + a_0 - b_0^2 \\ &= b_1^2 + \frac{a_0 - \sqrt{a_0^2 + a_1^2 + a_2^2 + a_3^2}}{2}. \end{aligned}$$

Equation (1.6) then becomes with $t = b_1^2$ and using (1.5)

$$\frac{a_1^2}{4t} \left(t + \frac{a_0 - \sqrt{a_0^2 + a_1^2 + a_2^2 + a_3^2}}{2} \right) = -\frac{(a_2^2 + a_3^2)}{4}$$

thus

$$t = \frac{a_1^2}{2} \frac{\left(\sqrt{a_0^2 + a_1^2 + a_2^2 + a_3^2} - a_0 \right)}{a_1^2 + a_2^2 + a_3^2},$$

hence b_1 (one proceeds similarly for b_2, b_3); finally, one obtains

$$b_0 = \frac{\varepsilon}{\sqrt{2}}\sqrt{\sqrt{a_0^2+a_1^2+a_2^2+a_3^2}+a_0} \qquad (\varepsilon = \pm 1),$$

$$b_1 = \frac{\varepsilon}{\sqrt{2}}\frac{a_1}{\sqrt{a_1^2+a_2^2+a_3^2}}\sqrt{\sqrt{a_0^2+a_1^2+a_2^2+a_3^2}-a_0},$$

$$b_2 = \frac{\varepsilon}{\sqrt{2}}\frac{a_2}{\sqrt{a_1^2+a_2^2+a_3^2}}\sqrt{\sqrt{a_0^2+a_1^2+a_2^2+a_3^2}-a_0},$$

$$b_2 = \frac{\varepsilon}{\sqrt{2}}\frac{a_3}{\sqrt{a_1^2+a_2^2+a_3^2}}\sqrt{\sqrt{a_0^2+a_1^2+a_2^2+a_3^2}-a_0}.$$

Example. Consider the quaternion $a = 1 + i + j + k$; find its square root b

$$b_0 = \frac{\varepsilon}{\sqrt{2}}\sqrt{3}, \qquad b_1 = b_2 = b_3 = \frac{\varepsilon}{\sqrt{6}},$$

$$\text{Answer: } b = \pm\frac{1}{\sqrt{2}}\left(\sqrt{3}+\frac{i}{\sqrt{3}}+\frac{j}{\sqrt{3}}+\frac{k}{\sqrt{3}}\right).$$

n^{th} roots

The nth root of a quaternion $a = r(\cos\varphi + \mathbf{u}\sin\varphi)$, where φ can always be chosen within the interval $[0, \pi]$ with an appropriate choice of $\mathbf{u}$, is obtained as follows [9].

1. Supposing $\sin\varphi \neq 0$, the equation $b^n = a$ with $b = R(\cos\theta + \mathbf{e}\sin\theta)$, $\theta \in [0, \pi]$, leads to

   $$R^n = r, \quad \cos n\theta = \cos\varphi, \quad \sin n\theta = \sin\varphi, \quad \mathbf{e} = \mathbf{u}$$

   and thus to

   $$R = r^{\frac{1}{n}}, \quad \theta = \frac{(\varphi + 2k\pi)}{n} \qquad (k = 0, 1, \ldots, n-1);$$

   finally, one has

   $$b = r^{\frac{1}{n}}\left[\cos\frac{(\varphi+2k\pi)}{n} + \mathbf{u}\sin\frac{(\varphi+2k\pi)}{n}\right] \qquad (k = 0, 1, \ldots, n-1).$$

2. When $\sin\varphi = 0$, the vector $\mathbf{e}$ in b is arbitrary. If $a > 0$, one has $\varphi = 0$ and thus $\theta = \frac{2\pi m}{n}$ $(m = 0, 1, \ldots, n-1)$. For $n = 2$, one obtains $\theta = 0, \pi$ and thus the real roots $\pm\sqrt{a}$. With $n > 2$, certain values of θ ($\neq 0$ or π) give nonreal roots, the vector $\mathbf{e}$ being arbitrary. With $a < 0$, one has $\varphi = \pi$, $\theta = \frac{(2m+1)\pi}{n}$ $(m = 0, 1, \ldots, n-1)$, certain values of $\theta \neq \pi$ give nonreal roots b , the vector $\mathbf{e}$ being arbitrary.

Example. Find the cubic root of

$$\begin{aligned} a &= 3+i+j+k \\ &= 2\sqrt{3}\left[\cos 30^\circ + \frac{(i+j+k)}{\sqrt{3}}\sin 30^\circ\right]; \end{aligned}$$

$$\begin{aligned} \text{Answer: } b &= \left(2\sqrt{3}\right)^{\frac{1}{3}}\left[\cos\theta + \frac{(i+j+k)}{\sqrt{3}}\sin\theta\right], \\ \theta &= 10^\circ,\ 130^\circ,\ 250^\circ. \end{aligned}$$

1.4.4 Other functions and representations of quaternions

The exponential e^a is defined by [30]

$$e^a = 1 + \frac{a}{1!} + \frac{a^2}{2!} + \frac{a^3}{3!} + \cdots$$

where a is an arbitrary quaternion. Furthermore, one defines

$$\cosh a = \frac{e^a + e^{-a}}{2} = 1 + \frac{a^2}{2!} + \frac{a^4}{4!} + \cdots + \frac{a^{2p}}{(2p)!},$$

$$\sinh a = \frac{e^a - e^{-a}}{2} = \frac{a}{1!} + \frac{a^3}{3!} + \frac{a^5}{5!} + \cdots + \frac{a^{2p+1}}{(2p+1))!},$$

and thus $e^a = \cosh a + \sinh a$. For an arbitrary quaternion a, let $U(V(a)) = \mathbf{u}$ ($\mathbf{u}^2 = -1$) be a unit vector, and therefore one has $\mathbf{u}a = a\mathbf{u}$; consequently, one can define the trigonometric functions

$$\cos a = \frac{e^{\mathbf{u}a} + e^{-\mathbf{u}a}}{2} = 1 - \frac{a^2}{2!} + \frac{a^4}{4!} + \cdots,$$

$$\sin a = \frac{e^{\mathbf{u}a} - e^{-\mathbf{u}a}}{2} = \frac{a}{1!} - \frac{a^3}{3!} + \frac{a^5}{5!} + \cdots.$$

Example. Let $a = \mathbf{u}\theta$ be a quaternion without a scalar part with $\mathbf{u} \in \operatorname{Vec}\mathbb{H}$, $\mathbf{u}^2 = -1$ and θ real; one has

$$\begin{aligned} \cosh \mathbf{u}\theta &= \cos\theta, \\ \sinh \mathbf{u}\theta &= u\sin\theta, \\ e^{\mathbf{u}\theta} &= \cos\theta + \mathbf{u}\sin\theta, \\ \cos \mathbf{u}\theta &= \cosh\theta, \\ \sin \mathbf{u}\theta &= u\sinh\theta. \end{aligned}$$

In particular, if $a = i\theta$,

$$\cosh i\theta = \cos\theta, \qquad \sinh i\theta = i\sin\theta,$$

$$\cos i\theta = \cosh\theta, \qquad \sin i\theta = i\sinh\theta.$$

One can represent a quaternion $a = a_0 + a_1 i + a_2 j + a_3 k$ by a 2×2 complex matrix (with i' being the usual complex imaginary)

$$A = \begin{bmatrix} a_0 + i'a_3 & -i'a_1 + a_2 \\ -i'a_1 - a_2 & a_0 - i'a_3 \end{bmatrix}$$

or by a 4×4 real matrix

$$A = \begin{bmatrix} a_0 & -a_1 & -a_2 & -a_3 \\ a_1 & a_0 & -a_3 & a_2 \\ a_2 & a_3 & a_0 & -a_1 \\ a_3 & -a_2 & a_1 & a_0 \end{bmatrix}.$$

The differential of a product of quaternions is given by

$$d(ab) = (da)b + a(db), \qquad a, b \in \mathbb{H},$$

the order of the factors having to be respected.

1.5 Classical vector calculus

1.5.1 Scalar product and vector product

Let $a, b, c \in \operatorname{Vec} \mathbb{H}$, be three quaternions without a scalar part, $a = a_1 i + a_2 j + a_3 k$, $b = b_1 i + b_2 j + b_3 k$, $c = c_1 i + c_2 j + c_3 k$. The norm of a is

$$|a| = \sqrt{a a_c} = \sqrt{a_1^2 + a_2^2 + a_3^2}$$

and

$$\begin{aligned} ab &= \frac{ab + ba}{2} + \frac{ab - ba}{2} \\ &= (-\mathbf{a} \cdot \mathbf{b}, \mathbf{a} \times \mathbf{b}). \end{aligned}$$

One defines respectively the scalar product and the vector product of two vectors a, b by

$$\begin{aligned} (a, b) \equiv \mathbf{a} \cdot \mathbf{b} &= -\frac{(ab + ba)}{2} = a_1 b_1 + a_2 b_2 + a_3 b_3, \\ a \times b \equiv \mathbf{a} \times \mathbf{b} &= \frac{(ab - ba)}{2} \\ &= (a_2 b_3 - a_3 b_2) i + (a_3 b_1 - a_1 b_3) j + (a_1 b_2 - a_2 b_1) k, \end{aligned}$$

with $ab = -(a, b) + a \times b$. Geometrically, one has

$$\begin{aligned} (a, b) &= |a|\,|b| \cos \alpha, \\ |a \times b| &= |a|\,|b| \sin \alpha, \end{aligned}$$

α being the angle between the two vectors a and b. Furthermore,

$$\begin{aligned}|ab|^2 = |a|^2 |b|^2 &= (ab)(ab)_c \\ &= [-(a,b) + a \times b][-(a,b) - a \times b] \\ &= (a,b)^2 - (a \times b)^2 \\ &= (a,b)^2 + |a \times b|^2\end{aligned}$$

which is coherent with the above geometrical expressions. One has the properties

$$\begin{aligned}(a,b) &= (b,a), \\ (\lambda a, b) &= \lambda(a,b), \qquad \lambda \in \mathbb{R}, \\ (a, b+c) &= (a,b) + (a,c), \\ a \times b &= -b \times a, \\ a \times \lambda b &= \lambda(a \times b), \\ a \times (b+c) &= a \times b + a \times c.\end{aligned}$$

1.5.2 Triple scalar and double vector products

The triple scalar product is defined by

$$\begin{aligned}(a, b \times c) &= -\frac{[a(b \times c) + (b \times c)a]}{2} \\ &= -\frac{[a(bc - cb) + (bc - cb)a]}{2}\end{aligned}$$

and satisfies the relations

$$\begin{aligned}(a, b \times c) &= (b, c \times a) = (c, a \times b), \\ (a, b \times c) &= -(c, b \times a)\end{aligned}$$

which are established using the relations

$$S(abc) = S(bca) = S(cab)$$

and

$$\begin{aligned}S(abc) &= S(abc)_c = -S(cba) = -S(acb), \\ S(abc) &= -S(bac).\end{aligned}$$

In particular

$$(a \times b, a) = (a \times b, b) = 0$$

which shows that $a \times b$ is orthogonal to a and b.

To derive the expression of the double vector product $a \times (b \times c)$, one can start from

$$\begin{aligned} abc &= a\left[-(b,c) + b \times c\right] \\ &= -a(b,c) - (a, b \times c) + a \times (b \times c) \end{aligned}$$

hence

$$V(abc) = -a(b,c) + a \times (b \times c);$$

since

$$\begin{aligned} V(abc) &= \frac{abc - (abc)_c}{2} \\ &= \frac{abc + cba}{2} \\ &= \frac{[abc + bac - bac - bca + bca + cba]}{2} \\ &= \frac{(ab+ba)c}{2} - \frac{b(ac+ca)}{2} + \frac{(bc+cb)a}{2} \\ &= -(a,b)c + b(a,c) - a(b,c) \end{aligned}$$

one obtains

$$a \times (b \times c) = b(a,c) - c(a,b).$$

Knowing that

$$\begin{aligned} (a \times b) \times c &= -c \times (a \times b) \\ &= -a(c,b) + b(c,a), \end{aligned}$$

one notices that the vector product is not associative. From the above relations, one then obtains (with $a, b, c, d \in \operatorname{Vec} \mathbb{H}$)

$$\begin{aligned} (a \times b, c \times d) &= (a,c)(b,d) - (a,d)(b,c), \\ (a \times b) \times (c \times d) &= c(a \times b, d) - d(a \times b, c). \end{aligned}$$

1.6 Exercises

E1-1 From the formulas $i^2 = j^2 = k^2 = ijk = -1$, deduce the multiplication table of the quaternion group knowing that the element -1 commutes with all elements of the group and that $(-1)i = -i$, $(-1)j = -j$, $(-1)k = -k$.
E1-2 Consider the quaternions $a = 1 + i$, $b = 4 - 3j$. Compute $|a|$, $|b|$, a^{-1}, b^{-1}, $a + b$, ab, ba. Give the polar form of a and b. Compute $\sqrt{a}$, $\sqrt{b}$, $a^{1/3}$.
E1-3 Solve in $x \in \mathbb{H}$, the equation $ax + xb = c$ ($a, b, c \in \mathbb{H}$, $aa_c \neq bb_c$).
N.A. : $a = 2i$, $b = j$, $c = k$, determine x.
E1-4 Solve in $x \in \mathbb{H}$, the equation $axb + cxd = e$ ($a, b, c, d, e \in \mathbb{H}$).
N.A. : $a = 2i$, $b = j$, $c = k$, $d = i$, $e = 3k$. Find x.
E1-5 Solve in $x \in \mathbb{H}$, the equation $x^2 = xa + bx$ ($a, b \in \mathbb{H}$).
N.A. : $a = -2j$, $b = -k$, determine x.

Chapter 2

Rotation groups SO(4) and SO(3)

In this chapter, the formulas of the rotation groups SO(4) and SO(3) are established from orthogonal symmetries. The crystallographic groups and Kepler's problem are then examined as applications of these groups.

2.1 Orthogonal groups O(4) and SO(4)

Consider two elements of a four-dimensional vector space $x = x_0 + x_1 i + x_2 j + x_3 k$, $y = y_0 + y_1 i + y_2 j + y_3 k \in \mathbb{H}$, and the scalar product

$$(x, x) = x x_c = x_0^2 + x_1^2 + x_2^2 + x_3^2.$$

One deduces from it

$$(x + y, x + y) = (x, x) + (y, y) + (x, y) + (y, x)$$

and postulating the relation $(x, y) = (y, x)$, one obtains

$$\begin{aligned}
(x, y) &= \frac{1}{2}\left[(x + y, x + y) - (x, x) - (y, y)\right] \\
&= \frac{1}{2}\left[(x + y)(x + y)_c - x x_c - y y_c\right] \\
&= \frac{1}{2}(x y_c + y x_c) \\
&= x_0 y_0 + x_1 y_1 + x_2 y_2 + x_3 y_3.
\end{aligned}$$

Two quaternions x, y are orthogonal if $(x, y) = 0$; a quaternion is unitary if $(x, x) = 1$. A hyperplane is defined by the relation $(a, x) = 0$ where a is a quaternion perpendicular to the hyperplane. The expression of a plane symmetry is obtained as follows.

Definition 2.1.1. The symmetric x' of x with respect to a hyperplane is obtained by drawing from x the perpendicular down to the hyperplane and by extending this perpendicular line by an equal length [11].

We shall assume that the hyperplanes go through the origin. The vector $x' - x$ is perpendicular to the hyperplane (and thus parallel to a) and $(x'+x)/2$ is perpendicular to a. Hence, the relations

$$x' = x + \lambda a, \qquad \lambda \in \mathbb{R},$$
$$\left(a, \frac{x'+x}{2}\right) = 0;$$

one then deduces

$$\left(a, x + \frac{\lambda a}{2}\right) = 0, \qquad \lambda = \frac{-2(a,x)}{(a,a)},$$
$$x' = x - \frac{2(a,x)a}{(a,a)}$$
$$= x - \frac{(ax_c + xa_c)a}{aa_c}$$
$$= -\frac{ax_c a}{aa_c}.$$

Theorem 2.1.2. *Any rotation of* $\mathrm{O}(n)$ *is the product of an even number* $\leq n$ *of symmetries; any inversion is the product of an odd number* $\leq n$ *of symmetries* [11].

A rotation is a proper transformation of a determinant equal to 1; an inversion is an improper transformation of a determinant equal to -1. In combining, in an even number, plane symmetries

$$x' = -mx_c m,$$

with $mm_c = 1$, one obtains the rotation group SO(4),

$$x' = axb,$$

with $a, b \in \mathbb{H}$ and $aa_c = bb_c = 1$. By including the inversions (odd number of symmetries) the expression of which are

$$x' = -ax_c b$$

with $a, b \in \mathbb{H}$ and $aa_c = bb_c = 1$, one obtains the orthogonal group O(4) with six parameters. This group, by definition, conserves the scalar product; indeed, for SO(4),

$$(x', y') = \frac{1}{2}\left[x'y'_c + y'x'_c\right]$$
$$= \frac{1}{2}\left[axbb_c ya_c + aybb_c xa_c\right]$$
$$= (x, y)$$

and for the improper rotations

$$\begin{aligned}(x', y') &= \frac{1}{2}\left[x'y'_c + y'x'_c\right] \\ &= \frac{1}{2}\left[ax_cbb_cya_c + ay_cbb_cxa_c\right] \\ &= (x, y).\end{aligned}$$

Any rotation of SO(4) can be written as a combination of a rotation of SO(3), examined below,

$$x = rxr_c, \qquad rr_c = 1$$

and a transformation

$$x' = axa, \qquad a \in \mathbb{H},\ aa_c = 1.$$

It is sufficient to resolve the equation

$$x' = fxg = arxr_ca \qquad (\text{or } r'a'x'a'r'_c)$$

with $f = ar$, $g = r_ca$ and $rr_c = aa_c = 1$. Writing the relations ([15], [16])

$$\begin{aligned}2a^2 &= 2fg, \\ a^2fg &= (fg)^2, \\ a^2(fg)_c &= 1,\end{aligned}$$

and adding, one obtains

$$a^2\left[2 + fg + (fg)_c\right] = (1 + fg)^2;$$

hence, the solution

$$a = \frac{\pm(1 + fg)}{|(1 + fg|}.$$

The rotation is given by

$$\begin{aligned}r &= a_cf \\ &= \pm\frac{(1 + g_cf_c)\,f}{|(1 + fg|} \\ &= \frac{\pm(f + g_c)}{|(1 + fg|}.\end{aligned}$$

One verifies that one has indeed the relations $aa_c = rr_c = 1$. One solves similarly the equations $f = r'a'$, $g = a'r'_c$ with the solutions

$$\begin{aligned}a' &= \frac{\pm(1 + gf)}{|(1 + gf|}, \\ r' &= \frac{\pm(f + g_c)}{|(1 + gf|},\end{aligned}$$

with $|(1 + fg| = |(1 + gf|$ since $S(gf) = S(fg)$.

2.2 Orthogonal groups O(3) and SO(3)

Consider the vectors $x = x_1 i + x_2 j + x_3 k$, $x' = x'_1 i + x'_2 j + x'_3 k \in \mathbb{H}$, constituting a subvectorspace of $\mathbb{H}$. A plane symmetry, in this subspace, is a particular case of the preceeding one and is expressed by

$$\begin{aligned} x' &= -a x_c a \\ &= a x_c a_c \end{aligned}$$

with $a \in \operatorname{Vec} \mathbb{H}$, $a a_c = 1$, $a_c = -a$. The improper rotations (odd number ≤ 3 of plane symmetries) are given by

$$x' = f x_c f_c$$

with $f \in \mathbb{H}$, $f f_c = 1$ and one has $x' x'_c = x x_c$. The SO(3) group is the set of proper rotations (even number ≤ 3 of symmetries). In combining two symmetries, one obtains

$$\begin{aligned} x'' &= -a x_c a, \\ x' &= -b x''_c b, \end{aligned}$$

hence

$$\begin{aligned} x' &= (b a_c) x (a_c b) \\ &= r x r_c \end{aligned}$$

with r $(= b a_c) \in \mathbb{H}$, $r_c = a b_c = a_c b$, $r r_c = 1$. The unitary quaternion r can be expressed in the form

$$r = \left(\cos\frac{\theta}{2} + \mathbf{u}\sin\frac{\theta}{2}\right)$$

where the unit vector $\mathbf{u}$ ($\mathbf{u}^2 = -1$) is the axis of rotation (going through the origin) and θ the angle of rotation of the vector x around $\mathbf{u}$ (θ is taken algebraically given the direction of $\mathbf{u}$ and using the right-handed screw rule). The conservation of the norm of x results from

$$x' x'_c = r x r_c r x_c r_c = x x_c.$$

Furthermore, if one considers the transformation

$$q' = r q r_c$$

with $q \in \mathbb{H}$, one has

$$S(q') = S(r q r_c) = S(r_c r q) = S(q)$$

which shows that the scalar part of the quaternion is not affected by the rotation. The set of proper and improper rotations constitute the group O(3). In developing the formula $x' = rxr_c$ with $x = \mathbf{x} \in \operatorname{Vec} \mathbb{H}$, one obtains

$$\begin{aligned} x' = \mathbf{x}' &= \left(\cos\frac{\theta}{2} + \mathbf{u}\sin\frac{\theta}{2}\right)\mathbf{x}\left(\cos\frac{\theta}{2} - \mathbf{u}\sin\frac{\theta}{2}\right) \\ &= \cos^2\frac{\theta}{2}\mathbf{x} + \sin^2\frac{\theta}{2}\left[(\mathbf{u}\cdot\mathbf{x})\mathbf{u} - (\mathbf{u}\times\mathbf{x})\mathbf{u}\right] + \mathbf{u}\times\mathbf{x}\sin\theta; \end{aligned}$$

furthermore

$$\begin{aligned} (\mathbf{u}\times\mathbf{x})\mathbf{u} &= -(\mathbf{u}\times\mathbf{x})\cdot\mathbf{u} + (\mathbf{u}\times\mathbf{x})\times\mathbf{u} \\ &= (\mathbf{u}\times\mathbf{x})\times\mathbf{u} \\ &= \mathbf{x} - (\mathbf{u}\cdot\mathbf{x})\mathbf{u}; \end{aligned}$$

hence, the classical formula [26, p. 165]

$$\mathbf{x}' = \mathbf{x}\cos\theta + \mathbf{u}(\mathbf{u}\cdot\mathbf{x})(1-\cos\theta) + \mathbf{u}\times\mathbf{x}\sin\theta.$$

In matrix form, this equation can be written $x' = Ax$ with

$$x = \begin{bmatrix} 0 \\ x_1 \\ x_2 \\ x_3 \end{bmatrix}, \qquad x' = \begin{bmatrix} 0 \\ x_1' \\ x_2' \\ x_3' \end{bmatrix},$$

$$A = \begin{bmatrix} \begin{pmatrix} u_1^2 \\ +(u_2^2+u_3^2)\cos\theta \end{pmatrix} & \begin{pmatrix} u_1u_2(1-\cos\theta) \\ -u_3\sin\theta \end{pmatrix} & \begin{pmatrix} u_1u_3(1-\cos\theta) \\ +u_2\sin\theta \end{pmatrix} \\ \begin{pmatrix} u_1u_2(1-\cos\theta) \\ +u_3\sin\theta \end{pmatrix} & \begin{pmatrix} u_2^2 \\ +(u_1^2+u_3^2)\cos\theta \end{pmatrix} & \begin{pmatrix} u_2u_3(1-\cos\theta) \\ -u_1\sin\theta \end{pmatrix} \\ \begin{pmatrix} u_1u_3(1-\cos\theta) \\ -u_2\sin\theta \end{pmatrix} & \begin{pmatrix} u_2u_3(1-\cos\theta) \\ +u_1\sin\theta \end{pmatrix} & \begin{pmatrix} u_3^2 \\ +(u_1^2+u_2^2)\cos\theta \end{pmatrix} \end{bmatrix};$$

one verifies that the matrix A is orthogonal (${}^tA = A^{-1}$). If one combines the rotation r_1 and the rotation r_2,

$$r_1 = \left(\cos\frac{\alpha}{2} + \mathbf{a}\sin\frac{\alpha}{2}\right), \qquad r_2 = \left(\cos\frac{\beta}{2} + \mathbf{b}\sin\frac{\beta}{2}\right)$$

one obtains

$$r = r_2 r_1 = \left(\cos\frac{\gamma}{2} + \mathbf{c}\sin\frac{\gamma}{2}\right)$$

with

$$\begin{aligned} \cos\frac{\gamma}{2} &= \cos\frac{\alpha}{2}\cos\frac{\beta}{2} - (\mathbf{a},\mathbf{b})\sin\frac{\alpha}{2}\sin\frac{\beta}{2}, \\ \mathbf{c}\sin\frac{\gamma}{2} &= \mathbf{a}\sin\frac{\alpha}{2}\cos\frac{\beta}{2} + \mathbf{b}\cos\frac{\alpha}{2}\sin\frac{\beta}{2} - (\mathbf{a}\times\mathbf{b})\sin\frac{\alpha}{2}\sin\frac{\beta}{2}, \end{aligned}$$

which yields the Rodriguez formula

$$\mathbf{c}\tan\frac{\gamma}{2} = \frac{\mathbf{a}\tan\frac{\alpha}{2} + \mathbf{b}\tan\frac{\beta}{2} - (\mathbf{a}\times\mathbf{b})\tan\frac{\alpha}{2}\tan\frac{\beta}{2}}{1 - (\mathbf{a},\mathbf{b})\tan\frac{\alpha}{2}\tan\frac{\beta}{2}}.$$

2.3 Crystallographic groups

If r belongs to a finite subgroup of real quaternions, the transformations $q' = rqr_c$ will constitute a subgroup of SO(3), with r and $-r$ generating the same rotation. The finite subgroups of real quaternions ([48], [44]) are of five types.

2.3.1 Double cyclic groups C_n (order $N = 2n$)

The elements of these groups are given by

$$\begin{aligned} r = \mathbf{u}^{\frac{2h}{n}} &= \left(\cos\frac{\pi}{2} + \mathbf{u}\sin\frac{\pi}{2}\right)^{\frac{2h}{n}} \\ &= \left(\cos\frac{\pi}{n} + \mathbf{u}\sin\frac{\pi}{n}\right)^{h} = b^h \end{aligned}$$

with $b = \cos\frac{\pi}{n} + \mathbf{u}\sin\frac{\pi}{n}$, the axis being oriented according to the unit vector $\mathbf{u}$, with $h = 1, \ldots, 2n$. If the rotation axis is oriented along Oz, one simply has

$$\begin{aligned} r = k^{\frac{2h}{n}} &= \left(\cos\frac{\pi}{2} + k\sin\frac{\pi}{2}\right)^{\frac{2h}{n}} \\ &= \left(\cos\frac{\pi}{n} + k\sin\frac{\pi}{n}\right)^{h}. \end{aligned}$$

Example. Double group C_3 ($N = 6$, rotation axis along Oz); the elements of the group are

$$r = b^h = \left(\cos\frac{\pi}{3} + k\sin\frac{\pi}{3}\right)^h, \qquad h = 1, \ldots, 6$$

or explicitly

$$\left\{ \begin{array}{c} b = \frac{1}{2}(1 + k\sqrt{3}),\ b^2 = \frac{1}{2}(-1 + k\sqrt{3}),\ b^3 = -1, \\ b^4 = -b,\ b^5 = -b^2,\ b^6 = 1 \end{array} \right\}.$$

2.3.2 Double dihedral groups D_n ($N = 4n$)

These groups are constituted by the elements

$$r = u^{\frac{2h}{n}} a^l = b^h a^l$$

with

$$u = \left(\cos\frac{\pi}{2} + \mathbf{u}\sin\frac{\pi}{2}\right), \qquad a = \left(\cos\frac{\pi}{2} + \mathbf{a}\sin\frac{\pi}{2}\right), \qquad b = u^{\frac{2}{n}},$$

where **u**, **a** are two perpendicular vectors and $S(au) = 0$, $a^2 = -1$, $h = 1, \ldots, 2n$, $l = 1, \ldots, 4$.

Examples. 1. Double group D_3 (order 12)

$$b = \left(\cos\frac{\pi}{3} + k\sin\frac{\pi}{3}\right) = \frac{1}{2}(1 + k\sqrt{3}),$$

$$a = i, \qquad ab = \frac{1}{2}(i - j\sqrt{3}), \qquad ba = \frac{1}{2}(i + j\sqrt{3});$$

the elements of the group are $\{\pm b, \pm b^2, \pm b^3, \pm a, \pm ab, \pm ba\}$.

2. Double group D_4 (order 16); writing

$$b = \left(\cos\frac{\pi}{4} + k\sin\frac{\pi}{4}\right), \qquad b^2 = (k), \qquad b^3 = \frac{1}{\sqrt{2}}(-1 + k),$$

$$a = (i), \qquad ab = \frac{1}{\sqrt{2}}(i - j), \qquad ba = \frac{1}{\sqrt{2}}(i + j),$$

$$ab^2 = (-j), \qquad a^2 = (-1), \qquad a^3 = (-i), \qquad a^4 = 1,$$

the group is constituted by the elements

$$\{\pm b,\ \pm b^2,\ \pm b^3,\ \pm b^4,\ \pm a,\ \pm ab,\ \pm ba,\ \pm ab^2\}.$$

2.3.3 Double tetrahedral group ($N = 24$)

This group is constitued by the 24 elements

$$\pm 1, \quad \pm i, \quad \pm j, \quad \pm k,$$
$$\frac{1}{2}(\pm 1 \pm i \pm j \pm k).$$

More explicitly, indicating the axes of multiple rotations, the group is composed of the elements

$$i^\alpha, \quad j^\alpha, \quad k^\alpha,$$
$$\left(\frac{1 + i + j + k}{2}\right)^\beta, \quad \left(\frac{1 - i - j + k}{2}\right)^\beta,$$
$$\left(\frac{1 + i - j - k}{2}\right)^\beta, \quad \left(\frac{1 - i + j - k}{2}\right)^\beta,$$

with $\alpha = 1, 2, 3, 4$, $\beta = 1, 2, 3, 4, 5, 6$ ($N = 24$).

Example. Consider the tetraeder having as vertices

$$\frac{i + j + k}{\sqrt{3}}, \qquad \frac{-i - j + k}{\sqrt{3}}, \qquad \frac{i - j - k}{\sqrt{3}}, \qquad \frac{-i + j - k}{\sqrt{3}},$$

as face centers

$$\frac{-(i+j+k)}{\sqrt{3}}, \qquad -\frac{(-i-j+k)}{\sqrt{3}}, \qquad -\frac{(i-j-k)}{\sqrt{3}}, \qquad -\frac{(-i+j-k)}{\sqrt{3}},$$

and as side centers

$$\pm i, \qquad \pm j, \qquad \pm k;$$

by taking for r the elements of the above group, the transformation $x' = rxr_c$ transforms the tetraeder in itself.

2.3.4 Double octahedral group ($N = 48$)

The group is composed by the 48 elements

$$\begin{aligned}
&\pm 1, \quad \pm i, \quad \pm j, \quad \pm k, \\
&\frac{1}{2}(\pm 1 \pm i \pm j \pm k), \\
&\frac{1}{\sqrt{2}}(\pm 1 \pm i), \quad \frac{1}{\sqrt{2}}(\pm 1 \pm j), \quad \frac{1}{\sqrt{2}}(\pm 1 \pm k), \\
&\frac{1}{\sqrt{2}}(\pm i \pm j), \quad \frac{1}{\sqrt{2}}(\pm j \pm k), \quad \frac{1}{\sqrt{2}}(\pm i \pm k).
\end{aligned}$$

Making explicit the axes of multiple rotations, the elements of the group are

$$\begin{aligned}
&\left(\frac{1+i}{\sqrt{2}}\right)^\alpha, \quad \left(\frac{1+j}{\sqrt{2}}\right)^\alpha, \quad \left(\frac{1+k}{\sqrt{2}}\right)^\alpha, \\
&\left(\frac{1+i+j+k}{2}\right)^\beta, \quad \left(\frac{1-i-j+k}{2}\right)^\beta, \\
&\left(\frac{1+i-j-k}{2}\right)^\beta, \quad \left(\frac{1-i+j-k}{2}\right)^\beta, \\
&\left(\frac{j+k}{\sqrt{2}}\right)^\gamma, \quad \left(\frac{i+k}{\sqrt{2}}\right)^\gamma, \quad \left(\frac{i+j}{\sqrt{2}}\right)^\gamma, \\
&\left(\frac{j-k}{\sqrt{2}}\right)^\gamma, \quad \left(\frac{-i+k}{\sqrt{2}}\right)^\gamma, \quad \left(\frac{i-j}{\sqrt{2}}\right)^\gamma,
\end{aligned}$$

with $\alpha = 1, \ldots, 8$, $\beta = 1, \ldots, 6$, $\gamma = 1, \ldots, 4$ ($N = 48$).

Example. Consider the octaeder having, in an orthonormal frame, for its 6 vertices the coordinates $\pm i$, $\pm j$, $\pm k$, and for the centers of the 8 faces

$$\pm\frac{i+j+k}{\sqrt{3}}, \qquad \pm\frac{-i-j+k}{\sqrt{3}}, \qquad \pm\frac{i-j-k}{\sqrt{3}}, \qquad \pm\frac{-i+j-k}{\sqrt{3}},$$

for the middle points of the 12 sides

$$\pm\frac{j \pm k}{\sqrt{2}}, \qquad \pm\frac{(\pm i + k)}{\sqrt{2}}, \qquad \pm\frac{i \pm j}{\sqrt{2}}.$$

The octaeder transforms into itself under the rotation $x' = rxr_c$, r being taken in the double octahedral group. The same property applies to the cube (dual of the octaeder) the 8 vertices of which are the centers of the faces of the above octaeder.

2.3.5 Double icosahedral group ($N = 120$)

The 120 elements of this group are

$$i^\alpha, \quad j^\alpha, \quad k^\alpha,$$

$$\left(\frac{i + m'j + mk}{2}\right)^\alpha, \quad \left(\frac{mi + j + m'k}{2}\right)^\alpha, \quad \left(\frac{m'i + mj + k}{2}\right)^\alpha,$$

$$\left(\frac{i - m'j + mk}{2}\right)^\alpha, \quad \left(\frac{-mi + j - m'k}{2}\right)^\alpha, \quad \left(\frac{m'i - mj + k}{2}\right)^\alpha,$$

$$\left(\frac{i + m'j - mk}{2}\right)^\alpha, \quad \left(\frac{mi + j - m'k}{2}\right)^\alpha, \quad \left(\frac{-m'i - mj + k}{2}\right)^\alpha,$$

$$\left(\frac{i - m'j - mk}{2}\right)^\alpha, \quad \left(\frac{-mi + j + m'k}{2}\right)^\alpha, \quad \left(\frac{-m'i + mj + k}{2}\right)^\alpha,$$

$$\left(\frac{1 + i + j + k}{2}\right)^\beta, \quad \left(\frac{1 - i - j + k}{2}\right)^\beta,$$

$$\left(\frac{1 + i - j - k}{2}\right)^\beta, \quad \left(\frac{1 - i + j - k}{2}\right)^\beta,$$

$$\left(\frac{1 + mj + m'k}{2}\right)^\beta, \quad \left(\frac{1 + m'i + mk}{2}\right)^\beta, \quad \left(\frac{1 + mi + m'j}{2}\right)^\beta,$$

$$\left(\frac{1 + mj - m'k}{2}\right)^\beta, \quad \left(\frac{1 - m'i + mk}{2}\right)^\beta, \quad \left(\frac{1 + mi - m'j}{2}\right)^\beta,$$

$$\left(\frac{m + m'j + k}{2}\right)^\gamma, \quad \left(\frac{m + i + m'k}{2}\right)^\gamma, \quad \left(\frac{m + m'i + j}{2}\right)^\gamma,$$

$$\left(\frac{m + m'j - k}{2}\right)^\gamma, \quad \left(\frac{m - i + m'k}{2}\right)^\gamma, \quad \left(\frac{m + m'i - j}{2}\right)^\gamma,$$

with $\alpha = 1, \ldots, 4$, $\beta = 1, \ldots, 6$, $\gamma = 1, \ldots, 10$ and

$$m = \frac{1 + \sqrt{5}}{2} = 2\cos 36^\circ,$$

$$m' = \frac{1 - \sqrt{5}}{2} = -2\cos 72^\circ.$$

One obtains another group, distinct from the first, by inverting m and m'.

Example. Consider the icosaeder having, in an orthonormal frame, for the coordinates of the 12 vertices

$$\pm\frac{m'j \pm k}{m'\sqrt{5}}, \qquad \pm\frac{(\pm i + m'k)}{m'\sqrt{5}}, \qquad \pm\frac{m'i \pm j}{m'\sqrt{5}};$$

for the centers of the 20 faces

$$\pm\frac{i+j+k}{\sqrt{3}}, \qquad \pm\frac{-i-j+k}{\sqrt{3}}, \qquad \pm\frac{i-j-k}{\sqrt{3}}, \qquad \pm\frac{-i+j-k}{\sqrt{3}},$$

$$\pm\frac{mj \pm m'k}{\sqrt{3}}, \qquad \pm\frac{(\pm m'i + mk)}{\sqrt{3}}, \qquad \pm\frac{mi \pm m'j}{\sqrt{3}};$$

for the middles of the 30 sides

$$\pm i, \qquad \pm j, \qquad \pm k, \qquad \pm\frac{i \pm m'j \pm mk}{2},$$

$$\pm\frac{(\pm mi + j \pm m'k)}{2}, \qquad \pm\frac{(\pm m'i \pm mj + k)}{2}.$$

The transformation $x' = rxr_c$ transforms the icosaeder into itself; the same is true for its dual, the dodecaeder (having 20 vertices, 12 faces and 30 sides) the vertices of which are the centers of the faces of the icosaeder.

The five groups (cyclic, dihedral, tetrahedral, octahedral, icosahedral) above, combined with the rotations $x' = rxr_c$ and the parity operator $x' = x_c = -x$, generate the set of the 32 crystallographic groups([44], [43]).

2.4 Infinitesimal transformations of SO(4)

Among the subgroups of SO(4) one has, in particular, the transformations

$$x' = rxr_c, \qquad x' = axa, \qquad x' = fx, \qquad x' = xg$$

with $r,\ a,\ f,\ g \in \mathbb{H}$, $rr_c = aa_c = ff_c = gg_c = 1$, $x = (x_0, \mathbf{x})$ and $x' = (x'_0, \mathbf{x}')$ $\in \mathbb{H}$. For an infinitesimal rotation of SO(3) of $d\theta$ around the unit vector $\mathbf{u} = u^1 i + u^2 j + u^3 k$, one has

$$r = \left(\cos\frac{d\theta}{2} + \mathbf{u}\sin\frac{d\theta}{2}\right) \simeq \left(1 + \mathbf{u}\frac{d\theta}{2}\right)$$

and

$$\begin{aligned} x' = rxr_c &= \left(1 + \mathbf{u}\frac{d\theta}{2}\right) x \left(1 - \mathbf{u}\frac{d\theta}{2}\right) \\ &= x + \frac{d\theta}{2}(\mathbf{u}x - x\mathbf{u}) = x + d\theta\, \mathbf{u} \times \mathbf{x}; \end{aligned}$$

hence

$$dx = x' - x = d\theta\, \mathbf{u} \times \mathbf{x}.$$

In matrix notation, one obtains

$$dx = d\theta u^i M_i x, \qquad i \in (1, 2, 3)$$

with

$$x = \begin{bmatrix} x_0 \\ x_1 \\ x_2 \\ x_3 \end{bmatrix}, \qquad x' = \begin{bmatrix} x'_0 \\ x'_1 \\ x'_2 \\ x'_3 \end{bmatrix},$$

and

$$M_1 = \begin{bmatrix} 0 & 0 & 0 & 0 \\ 0 & 0 & 0 & 0 \\ 0 & 0 & 0 & -1 \\ 0 & 0 & 1 & 0 \end{bmatrix}, \quad M_2 = \begin{bmatrix} 0 & 0 & 0 & 0 \\ 0 & 0 & 0 & 1 \\ 0 & 0 & 0 & 0 \\ 0 & -1 & 0 & 0 \end{bmatrix}, \quad M_3 = \begin{bmatrix} 0 & 0 & 0 & 0 \\ 0 & 0 & -1 & 0 \\ 0 & 1 & 0 & 0 \\ 0 & 0 & 0 & 0 \end{bmatrix}.$$

Introducing the commutator of two matrices A and B,

$$[A, B] = AB - BA,$$

one verifies the relations

$$\begin{aligned} [M_1, M_2] &= M_3, \\ [M_1, M_3] &= -M_2, \\ [M_2, M_3] &= M_1, \end{aligned}$$

or

$$[M_i, M_j] = \epsilon_{ijk} M_k. \tag{2.1}$$

Concerning the subgroup $x' = axa$, with $a \in \mathbb{H}$ and $aa_c = 1$, one has for the infinitesimal transformation

$$a = \left(\cos\frac{d\beta}{2} + \mathbf{v}\sin\frac{d\beta}{2}\right) \simeq \left(1 + \mathbf{v}\frac{d\beta}{2}\right);$$

hence

$$\begin{aligned} x' = axa &= \left(1 + \mathbf{v}\frac{d\beta}{2}\right) x \left(1 + \mathbf{v}\frac{d\beta}{2}\right) \\ &= x + \frac{d\beta}{2}\left(\mathbf{v}x + x\mathbf{v}\right) = x + d\beta\left(-\mathbf{v}\cdot\mathbf{x} + x_0\mathbf{v}\right) \end{aligned}$$

and

$$dx = x' - x = d\beta\left(-\mathbf{v}\cdot\mathbf{x} + x_0\mathbf{v}\right).$$

In matrix notation, one obtains

$$dx = d\beta v^i N_i x, \qquad i \in (1,2,3)$$

with

$$x = \begin{bmatrix} x_0 \\ x_1 \\ x_2 \\ x_3 \end{bmatrix}, \qquad x' = \begin{bmatrix} x'_0 \\ x'_1 \\ x'_2 \\ x'_3 \end{bmatrix},$$

$$N_1 = \begin{bmatrix} 0 & -1 & 0 & 0 \\ 1 & 0 & 0 & 0 \\ 0 & 0 & 0 & 0 \\ 0 & 0 & 0 & 0 \end{bmatrix}, \quad N_2 = \begin{bmatrix} 0 & 0 & -1 & 0 \\ 0 & 0 & 0 & 0 \\ 1 & 0 & 0 & 0 \\ 0 & 0 & 0 & 0 \end{bmatrix}, \quad N_3 = \begin{bmatrix} 0 & 0 & 0 & -1 \\ 0 & 0 & 0 & 0 \\ 0 & 0 & 0 & 0 \\ 1 & 0 & 0 & 0 \end{bmatrix};$$

the matrices N_i satisfy the relations

$$\begin{aligned} [N_1, N_2] &= M_3, \\ [N_1, N_3] &= -M_2, \\ [N_2, N_3] &= M_1, \end{aligned}$$

or, more concisely

$$[N_i, N_j] = \epsilon_{ijk} M_k \tag{2.2}$$

where M_i are the matrices defined previously; furthermore

$$[N_i, M_j] = \epsilon_{ijk} N_k. \tag{2.3}$$

Concerning the subgroup $x' = fx$, one obtains with $f \simeq 1 + \frac{d\alpha}{2}\mathbf{f}$, $\mathbf{f} = f_1 i + f_2 j + f_3 k$ (unit vector)

$$dx = x' - x = \frac{d\alpha}{2}\left[-\mathbf{f}\cdot\mathbf{x} + x_0\mathbf{f} + \mathbf{f}\times\mathbf{x}\right].$$

In matrix notation, one has

$$dx = d\alpha f^i F_i x$$

with

$$F_1 = \frac{1}{2}\begin{bmatrix} 0 & -1 & 0 & 0 \\ 1 & 0 & 0 & 0 \\ 0 & 0 & 0 & -1 \\ 0 & 0 & 1 & 0 \end{bmatrix}, \qquad F_2 = \frac{1}{2}\begin{bmatrix} 0 & 0 & -1 & 0 \\ 0 & 0 & 0 & 1 \\ 1 & 0 & 0 & 0 \\ 0 & -1 & 0 & 0 \end{bmatrix},$$

$$F_3 = \frac{1}{2}\begin{bmatrix} 0 & 0 & 0 & -1 \\ 0 & 0 & -1 & 0 \\ 0 & 1 & 0 & 0 \\ 1 & 0 & 0 & 0 \end{bmatrix}$$

and

$$[F_1, F_2] = F_3,$$
$$[F_1, F_3] = -F_2,$$
$$[F_2, F_3] = F_1,$$

or

$$[F_i, F_j] = \epsilon_{ijk} F_k.$$

As to the subgroup $x' = xg$, with $g \simeq 1 + \frac{d\beta}{2}\mathbf{g}$, one proceeds similarly and obtains

$$dx = x' - x = \frac{d\beta}{2}\left[-\mathbf{g}\cdot\mathbf{x} + x_0\mathbf{g} - \mathbf{g}\times\mathbf{x}\right].$$

In matrix notation,

$$dx = d\beta g^i G_i x$$

with

$$G_1 = \frac{1}{2}\begin{bmatrix} 0 & -1 & 0 & 0 \\ 1 & 0 & 0 & 0 \\ 0 & 0 & 0 & 1 \\ 0 & 0 & -1 & 0 \end{bmatrix}, \qquad G_2 = \frac{1}{2}\begin{bmatrix} 0 & 0 & -1 & 0 \\ 0 & 0 & 0 & -1 \\ 1 & 0 & 0 & 0 \\ 0 & 1 & 0 & 0 \end{bmatrix},$$

$$G_3 = \frac{1}{2}\begin{bmatrix} 0 & 0 & 0 & -1 \\ 0 & 0 & 1 & 0 \\ 0 & -1 & 0 & 0 \\ 1 & 0 & 0 & 0 \end{bmatrix}$$

and

$$[G_1, G_2] = -G_3,$$
$$[G_1, G_3] = G_2,$$
$$[G_2, G_3] = -G_1,$$

or

$$[G_i, G_j] = -\epsilon_{ijk} G_k.$$

Furthermore, one has

$$[F_i, G_j] = 0, \qquad i, j = 1, 2, 3$$

the matrices M_i, N_i, F_i, G_i satisfying the relations

$$M_i = F_i - G_i, \qquad N_i = F_i + G_i.$$

2.5 Symmetries and invariants: Kepler's problem

Let us consider the motion of a particle of mass m, of momentum $\mathbf{p}$ and gravitating at the distance $\mathbf{r}$ of a mass M. The Hamiltonian is given by

$$H = \frac{p^2}{2m} - \frac{k}{r}, \qquad k = GMm$$

and the equations of motion are

$$\dot{q}_i = \frac{\partial H}{\partial p_i}, \qquad \dot{p}_i = \frac{-\partial H}{\partial q_i}.$$

Let $F(q_i, p_i, t)$ be a physical quantity of the motion; one has

$$\begin{aligned}\frac{dF}{dt} &= \frac{\partial F}{\partial t} + \frac{\partial F}{\partial q_i}\frac{\partial q_i}{\partial t} + \frac{\partial F}{\partial p_i}\frac{\partial p_i}{\partial t} \\ &= \frac{\partial F}{\partial t} + \left[\frac{\partial F}{\partial q_i}\frac{\partial H_i}{\partial p_i} - \frac{\partial F}{\partial p_i}\frac{\partial H}{\partial q_i}\right].\end{aligned}$$

Introducing Poisson's bracket of two functions u and v

$$[u, v] = \frac{\partial u}{\partial q_i}\frac{\partial v}{\partial p_i} - \frac{\partial u}{\partial p_i}\frac{\partial v}{\partial q_i}$$

one obtains

$$\frac{dF}{dt} = \frac{\partial F}{\partial t} + [F, H].$$

If $[F, H] = 0$ and $\frac{\partial F}{\partial t} = 0$, one has $\frac{dF}{dt} = 0$, F is then an invariant of the motion. The angular momentum $\mathbf{L} = \mathbf{r} \times \mathbf{p}$ and the Laplace-Runge-Lenz vector

$$\mathbf{A} = \mathbf{p} \times \mathbf{L} - \frac{km\mathbf{r}}{r}$$

satisfy the equations

$$\begin{aligned}[][\mathbf{L}, H] &= 0, \\ [\mathbf{A}, H] &= 0,\end{aligned}$$

and thus are invariants. Let us consider a bound motion (with a total negative energy $E < 0$) and introduce the vector

$$\mathbf{D} = \frac{\mathbf{A}}{\sqrt{-2mE}}$$

with

$$E = \frac{p^2}{2m} - \frac{k}{r}$$

verifying the Poisson bracket $[\mathbf{D}, H] = 0$; one has the relations

$$[L_i, L_j] = \epsilon_{ijk} L_k, \tag{2.4}$$

$$[D_i, D_j] = \epsilon_{ijk} L_k, \tag{2.5}$$

$$[D_i, L_j] = \epsilon_{ijk} D_k. \tag{2.6}$$

The relations (2.4), (2.5), (2.6) are respectively the same as those (2.1), (2.2), (2.3) concerning the infinitesimal transformations of SO(3) and the transformation $q' = aqa$ of SO(4), which indicates that the symmetry group of the problem is SO(4). To see it more explicitly, let us develop $\mathbf{A}$ in the form

$$\mathbf{A} = \left[\mathbf{r}\left(p^2 - \frac{km}{r}\right) - \mathbf{p}\,(\mathbf{p}\cdot\mathbf{r})\right]$$

with

$$\mathbf{D} = \frac{\mathbf{A}}{\sqrt{-2mE}} = \mathbf{r}p_0 - \mathbf{p}r_0$$

and

$$p_0 = \frac{p^2 - \frac{km}{r}}{\sqrt{-2mE}}, \qquad r_0 = \frac{\mathbf{p}\cdot\mathbf{r}}{\sqrt{-2mE}}.$$

One verifies immediately that $\mathbf{D}\cdot\mathbf{L} = 0$; furthermore,

$$(\mathbf{L})^2 = (\mathbf{r}\times\mathbf{p})^2 = r^2p^2 - (\mathbf{r}\cdot\mathbf{p})^2,$$

hence

$$(\mathbf{D})^2 + (\mathbf{L})^2 = \frac{k^2m^2}{-2mE}$$

or

$$H = \frac{-k^2m^2}{2\left[(\mathbf{D})^2 + (\mathbf{L})^2\right]}.$$

Introducing two quaternions

$$r = (r_0, \mathbf{r}), \qquad p = (p_0, \mathbf{p})$$

and the quaternion

$$K = r \wedge p = \frac{1}{2}(rp_c - pr_c) = (0, \mathbf{D} - \mathbf{L})$$

one has

$$KK_c = |0, \mathbf{D} - \mathbf{L}|^2 = \mathbf{D}^2 + \mathbf{L}^2,$$

and thus

$$H = \frac{-k^2m}{2\,|r\wedge p|^2}$$

which shows explicitly the invariance of the Hamiltonian with respect to a transformation of SO(4) of the type $K' = aKb$ ($a, b \in \mathbb{H}$, $aa_c = bb_c = 1$) leading to $K'K'_c = KK_c$.

2.6 Exercises

E2-1 From the general formula

$$x' = rxr_c, \qquad r = \left(\cos\frac{\theta}{2} + \mathbf{u}\sin\frac{\theta}{2}\right)$$

$x \in \mathrm{Vec}\,\mathbb{H}$, give in an orthonormal direct frame the matrix representation $X' = AX$ of a rotation of angle α around the Ox axis of a point $M(x, y, z)$, of a rotation of angle β around the Oy axis, of a rotation of angle γ around the Oz axis. Show that the matrices are orthogonal $A^{-1} = A^t$, $\det A = 1$ (A^t : transposed matrix of A).

E2-2 Consider the relation $A = rA'r_c$ where A are the components of a vector with respect to an orthonormal frame at rest and A' its components with respect to a mobile orthonormal frame. Show that $dA = rDA'r_c$ where DA' is the covariant differential with

$$DA' = dA' + d\Omega' \times A', \qquad d\Omega' = 2r_c dr$$

(dA' is the differential with respect to the components only). What does

$$\frac{d\Omega'}{dt} = 2r_c\frac{dr}{dt}$$

represent?

E2-3 Consider the relations

$$A = rA'r_c = gA''g_c$$

with

$$\begin{aligned} dA &= rDA'r_c = gDA''g_c, \\ DA' &= dA' + d\Omega' \times A' \qquad (d\Omega' = 2r_c dr), \\ DA'' &= dA'' + d\Omega'' \times A'' \qquad (d\Omega'' = 2g_c dg) \end{aligned}$$

(A represents the components of a vector with respect to an orthonormal basis at rest, A', A'' the components with respect to mobile orthonormal bases.). Find the relation between $d\Omega''$ and $d\Omega'$.

E2-4 Using the covariant derivative, express the velocity and the acceleration in polar coordinates (ρ, θ) and in cylindrical coordinates (ρ, θ, z).

E2-5 Express r in the relation $X = rX'r_c$ for spherical coordinates with $X = xi + yj + zk$ and $X' = \rho i$. Determine $d\Omega' = 2r_c dr$ and $d\Omega = 2(dr)r_c$. Express the basis vectors e_i in the basis at rest. Find the velocity and the acceleration in the mobile basis.

E2-6 Consider the Euler basis (O, x', y', z') obtained via the following successive rotations (Figure 2.1). A first rotation of angle α (precession angle) around k transforms the basis i, j, k into the basis i', j', k'. A second rotation of angle β (nutation angle) around the vector i' transforms i', j', k' into i'', j'', k''. A third rotation of angle γ (proper rotation angle) around the vector k'' transforms the basis i'', j'', k'' into the basis $e_1, e_2, e_3 = k''$. Give the quaternion r of the rotation $X = rX'r_c$. Determine

$$\omega' = 2r_c \frac{dr}{dt}, \qquad \omega = 2\frac{dr}{dt}r_c.$$

Give the components of the basis vectors e_i.

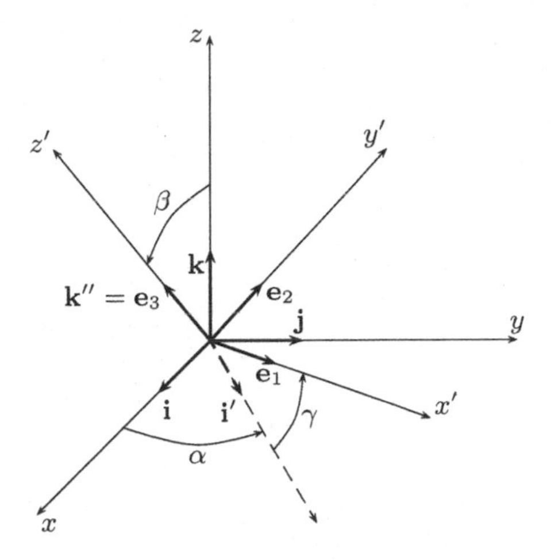

Figure 2.1: Euler's angles: α is the angle of precession, β the angle of nutation and γ the angle of proper rotation.

Chapter 3

Complex quaternions

From the very beginning of special relativity, complex quaternions have been used to formulate that theory [45]. This chapter establishes the expression of the Lorentz group using complex quaternions and gives a few applications. Complex quaternions constitute a natural transition towards the Clifford algebra $\mathbb{H} \otimes \mathbb{H}$.

3.1 Algebra of complex quaternions $\mathbb{H}(\mathbb{C})$

A complex quaternion is a quaternion $a = a_0 + a_1 i + a_2 j + a_3 k$ having complex coefficients. Such a quaternion can be represented by the matrix

$$A = \begin{bmatrix} a_0 + i' a_3 & -i' a_1 + a_2 \\ -i' a_1 - a_2 & a_0 - i' a_3 \end{bmatrix}$$

with the basis

$$1 = \begin{bmatrix} 1 & 0 \\ 0 & 1 \end{bmatrix}, \quad i = \begin{bmatrix} 0 & -i' \\ -i' & 0 \end{bmatrix}, \quad j = \begin{bmatrix} 0 & 1 \\ -1 & 0 \end{bmatrix}, \quad k = \begin{bmatrix} i' & 0 \\ 0 & -i' \end{bmatrix}$$

where i' is the usual complex imaginary and $a_i \in \mathbb{C}$. The algebra of complex quaternions $\mathbb{H}(\mathbb{C})$ is isomorphic to 2×2 matrices over $\mathbb{C}$ and has zero divisors; indeed, the relation

$$(1 + i'k)(1 - i'k) = 1 - (i')^2 k^2 = 0$$

shows that the product of two complex quaternions can be equal to zero without one of the complex quaternions being equal to zero.

3.2 Lorentz groups O(1, 3) and SO(1, 3)

3.2.1 Metric

With the advent of the special theory of relativity, space and time have been united into a four-dimensional pseudoeuclidean spacetime with the relativistic invariant

$$c^2t^2 - (x^1)^2 - (x^2)^2 - (x^3)^2 .$$

Minkowski had the idea of using complex quaternions

$$x = (i'ct, x^1, x^2, x^3) = (i'ct + \mathbf{x})$$

with the invariant

$$xx_c = -c^2t^2 + (x^1)^2 + (x^2)^2 + (x^3)^2 .$$

To get the signature $(+---)$, we shall use complex quaternions in the form

$$x = (ct, i'x^1, i'x^2, i'x^3) = (ct + i'\mathbf{x})$$

which we shall call minkowskian quaternions or minquats with the invariant

$$xx_c = c^2t^2 - (x^1)^2 - (x^2)^2 - (x^3)^2 .$$

A minquat is a complex quaternion such that $x_c^* = x$ where $*$ is the complex conjugation.

3.2.2 Plane symmetry

Let us consider two minquats $x = (x^0 + i'\mathbf{x})$, $y = (y^0 + i'\mathbf{y})$ with the scalar product

$$xx_c = (x^0)^2 - (x^1)^2 - (x^2)^2 - (x^3)^2 .$$

One has

$$(x+y, x+y) = (x,x) + (y,y) + (x,y) + (y,x);$$

postulating the property $(x,y) = (y,x)$ one obtains

$$\begin{aligned}(x,y) &= \frac{1}{2}\left[(x+y,x+y) - (x,x) - (y,y)\right]\\ &= \frac{1}{2}\left[(x+y)(x+y)_c - xx_c - yy_c\right]\\ &= \frac{1}{2}(xy_c + yx_c)\\ &= x^0y^0 - x^1y^1 - x^2y^2 - x^3y^3 .\end{aligned}$$

Two minquats x, y are said to be orthogonal if $(x,y) = 0$. A minquat x is timelike if $xx_c > 0$ and unitary if $xx_c = 1$; a minquat x is spacelike if $xx_c < 0$ and unitary if $xx_c = -1$. A hyperplane is defined by the relation $(a,x) = 0$ where a is a minquat perpendicular to the hyperplane.

Definition 3.2.1. The symmetric x' of x with respect to a hyperplane x is obtained by drawing the perpendicular to the hyperplane and by extending this perpendicular by an equal length [11].

The hyperplanes are supposed to go through the origin. The vector $x' - x$ is perpendicular to the hyperplane and thus parallel to a; furthermore, the vector $(x' + x)/2$ is perpendicular to a. One has the relations

$$x' = x + \lambda a, \qquad \lambda \in \mathbb{R},$$
$$\left(a, \frac{x' + x}{2}\right) = 0;$$

hence,

$$\left(a, x + \frac{\lambda a}{2}\right) = 0,$$
$$\lambda = \frac{-2(a, x)}{(a, a)},$$
$$x' = x - \frac{2(a, x)a}{(a, a)} = x - \frac{(ax_c + xa_c)a}{aa_c},$$
$$x' = -\frac{ax_c a}{aa_c}.$$

One shall distinguish time symmetries (with $aa_c = 1$)

$$x' = -ax_c a$$

from space symmetries ($aa_c = -1$)

$$x' = ax_c a.$$

3.2.3 Groups O(1, 3) and SO(1, 3)

Definition 3.2.2. The pseudo-orthogonal group O(1, 3) is the group of linear operators which leave invariant the quadratic form

$$(x, y) = x^0 y^0 - x^1 y^1 - x^2 y^2 - x^3 y^3$$

with $x = (x^0 + i'\mathbf{x})$, $y = (y^0 + i'\mathbf{y})$.

Theorem 3.2.3. *Any rotation of* O(1, 3) *is the product of an even number* ≤ 4 *of symmetries; any inversion is the product of an odd number* ≤ 4 *of symmetries* [11].

A rotation is a proper transformation of a determinant equal to 1; an inversion is an improper transformation of a determinant equal to -1.

Proper orthochronous Lorentz transformation

Consider the transformation obtained by combining an even number of time symmetries and an even number of space symmetries. These transformations of determinant equal to +1 constitute a subgroup of $O(1,3)$, the special orthogonal group $SO(1,3)$. Let us take for example two time symmetries (minquats f and g) followed by two space symmetries (minquats m and n). One has

$$\begin{aligned} x' &= n\left[m\left[g\left(fx_c f\right)_c g\right]_c m\right]_c n \\ &= (nm_c g f_c)x(f_c g m_c n); \end{aligned}$$

writing $a = nm_c g f_c$, one obtains

$$\begin{aligned} a_c &= f g_c m n_c, \qquad aa_c = 1 \\ a_c^* &= f^* g_c^* m^* n_c^* \\ &= f_c g m_c n; \end{aligned}$$

hence, the formula is valid in the general case,

$$x' = axa_c^*$$

with $aa_c = 1$, $a \in \mathbb{H}(\mathbb{C})$.

Other Lorentz transformations

Call n the number of time symmetries and p the number of space symmetries. In combining these symmetries, one obtains the following Lorentz transformations L:

1. n even, p odd, orthochronous, improper Lorentz transformation ($\det L = -1$)

$$x' = ax_c a_c^* \qquad (aa_c = -1);$$

2. n odd, p odd, antichronous, proper Lorentz transformation($\det L = 1$)

$$x' = -axa_c^* \qquad (aa_c = -1);$$

3. n odd, p even, antichronous, improper Lorentz transformation ($\det L = -1$)

$$x' = -ax_c a_c^* \qquad (aa_c = 1).$$

Group $O(1,3)$: summarizing table

The whole set of Lorentz transformations is given in the table below where n is the number of time symmetries and p the number of space symmetries [49].

	Rotation ($\det L = 1$)	Inversion ($\det L = -1$)
orthochronous	n even, p even $x' = axa_c^*$ $(aa_c = 1)$	n even, p odd $x' = ax_c a_c^*$ $(aa_c = -1)$
antichronous	n odd, p odd $x' = -axa_c^*$ $(aa_c = -1)$	n odd, p even $x' = -ax_c a_c^*$ $(aa_c = 1)$

3.3 Orthochronous, proper Lorentz group

3.3.1 Properties

The proper orthochronous Lorentz transformation

$$x' = axa_c^*$$

with $aa_c = 1, x = (ct + i'\mathbf{x}), x' = (ct' + i'\mathbf{x}')$ conserves, by definition, the norm

$$\begin{aligned} xx_c' &= (axa_c^*)(a^* x_c a_c) \\ &= xx_c \end{aligned}$$

and the minquat type of x

$$\begin{aligned} x_c'^* &= (axa_c^*)_c^* = ax_c^* a_c^* \\ &= axa_c^* = x'. \end{aligned}$$

The composition of two transformations satisfies the rule

$$\begin{aligned} x' &= a_2(a_1 x a_{1c}^*) a_{2c}^* \\ &= a_3 x a_{3c}^* \end{aligned}$$

with $a_3 = a_2 a_1$, $a_3 a_{3c} = 1$, $a_3 \in \mathbb{H}(\mathbb{C})$. A (three-dimensional) rotation is given by the formula

$$x' = rxr_c$$

with $r = \left(\cos\frac{\theta}{2} + \mathbf{u}\sin\frac{\theta}{2}\right), rr_c = 1, r \in \mathbb{H}$. A pure Lorentz transformation (without rotation) corresponds to the transformation

$$x' = bxb_c^*$$

where $b = \left(\cosh\frac{\varphi}{2} + i'\mathbf{v}\sinh\frac{\varphi}{2}\right)$ is a minquat such that $bb_c = 1$ and $i'\mathbf{v}$ is a unitary space-vector $(i'\mathbf{v}, i'\mathbf{v}) = -1$. A general transformation (proper, orthochronous) is obtained by combining a rotation and a pure Lorentz transformation. This can be done in two ways,

$$x' = b(rxr_c)b_c^*,$$
$$x'' = r(bxb_c^*)r_c,$$

br being in general distinct from rb. Reciprocally, a general Lorentz transformation can be decomposed into a pure Lorentz transformation and a rotation. The problem simply consists to resolve the equation $a = br$ (or $r'b'$) where b is a unitary minquat ($bb_c = 1$) and r a real unitary quaternion ($rr_c = 1$). The equation $a = br$ is solved in the following way ([15], [16]). Since

$$a_c^* = r_c^* b_c^* = r_c b$$

one obtains $aa_c^* = b^2$; let us write $d = aa_c^*$, with $dd_c = aa_c^* a^* a_c = 1$, hence, the equations

$$2b^2 = 2d,$$
$$b^2 d = d^2,$$
$$b^2 d_c = 1$$

and their sum

$$b^2(2 + d + d_c) = (1 + d)^2.$$

The solution therefore is

$$b = \frac{\pm(1 + d)}{|1 + d|}$$
$$= \frac{\pm(1 + aa_c^*)}{|1 + aa_c^*|}$$

with $|1 + d| = \sqrt{(1 + d)(1 + d)_c}$. The rotation is given by

$$r = b_c a$$
$$= \frac{\pm(a + a^*)}{|1 + aa_c^*|}.$$

Finally, one verifies that this is indeed a solution. For the equation $a = r'b'$, one finds in a similar way the solution

$$b' = \frac{\pm(1 + a_c^* a)}{|1 + a_c^* a|},$$
$$r' = \frac{\pm(a + a^*)}{|1 + a_c^* a|}$$

with $|1 + a_c^* a| = |1 + a a_c^*|$ since $S(aa_c^*) = S(a_c^* a)$. One observes that in both cases ($a = br$ or $r'b'$) the rotation is the same. The problem of the decomposition of a Lorentz transformation into a pure Lorentz transformation and a rotation is thus solved in the most general case. As an immediate application, consider the combination of two pure Lorentz transformations b_1, b_2. The quaternion $b = b_2 b_1$ will in general be a complex quaternion (and not a minquat) and will be written $a = br$. The resulting Lorentz transformation will thus contain a rotation; this is the principle of the Thomas precession ([52], [51]).

3.3.2 Infinitesimal transformations of SO(1, 3)

Consider the pure Lorentz transformation

$$x' = bxb_c^*$$

with $x = (ct + i'\mathbf{x})$, $x' = (ct' + i'\mathbf{x}')$, $b = \left(\cosh\frac{\varphi}{2} + i'\mathbf{v}\sinh\frac{\varphi}{2}\right)$ and $i'\mathbf{v}$ a spacelike unitary minquat. For an infinitesimal transformation,

$$\begin{aligned} b &\simeq \left(\cosh\frac{d\varphi}{2} + i'\mathbf{v}\sinh\frac{d\varphi}{2}\right) \\ &= \left(1 + i'\mathbf{v}\frac{d\varphi}{2}\right), \end{aligned}$$

hence

$$\begin{aligned} x' &= bxb_c^* \\ &= \left(1 + i'\mathbf{v}\frac{d\varphi}{2}\right) x \left(1 + i'\mathbf{v}\frac{d\varphi}{2}\right) \\ &= x + i'\frac{d\varphi}{2}\left(\mathbf{v}x + x\mathbf{v}\right); \end{aligned}$$

consequently,

$$\begin{aligned} dx &= x' - x \\ &= d\varphi\left(\mathbf{v}\cdot\mathbf{x} + i'\mathbf{v}x_0\right). \end{aligned}$$

Using real matrices,

$$X = \begin{bmatrix} x^0 = ct \\ x^1 \\ x^2 \\ x^3 \end{bmatrix}, \qquad X' = \begin{bmatrix} x'^0 = ct' \\ x'^1 \\ x'^2 \\ x'^3 \end{bmatrix}$$

one can write

$$\begin{aligned} dX &= X' - X \\ &= d\varphi v^i K_i X \end{aligned}$$

with

$$K_1 = \begin{bmatrix} 0 & 1 & 0 & 0 \\ 1 & 0 & 0 & 0 \\ 0 & 0 & 0 & 0 \\ 0 & 0 & 0 & 0 \end{bmatrix}, \quad K_2 = \begin{bmatrix} 0 & 0 & 1 & 0 \\ 0 & 0 & 0 & 0 \\ 1 & 0 & 0 & 0 \\ 0 & 0 & 0 & 0 \end{bmatrix}, \quad K_3 = \begin{bmatrix} 0 & 0 & 0 & 1 \\ 0 & 0 & 0 & 0 \\ 0 & 0 & 0 & 0 \\ 1 & 0 & 0 & 0 \end{bmatrix}.$$

The matrices K_i satisfy the relations

$$[K_i, K_j] = -\epsilon_{ijk} M_k, \tag{3.1}$$

$$[K_i, M_j] = \epsilon_{ijk} K_k, \tag{3.2}$$

where M_i are the matrices defined for the infinitesimal transformations of SO(3) in Chapter 2,

$$M_1 = \begin{bmatrix} 0 & 0 & 0 & 0 \\ 0 & 0 & 0 & 0 \\ 0 & 0 & 0 & -1 \\ 0 & 0 & 1 & 0 \end{bmatrix}, \quad M_2 = \begin{bmatrix} 0 & 0 & 0 & 0 \\ 0 & 0 & 0 & 1 \\ 0 & 0 & 0 & 0 \\ 0 & -1 & 0 & 0 \end{bmatrix}, \quad M_3 = \begin{bmatrix} 0 & 0 & 0 & 0 \\ 0 & 0 & -1 & 0 \\ 0 & 1 & 0 & 0 \\ 0 & 0 & 0 & 0 \end{bmatrix}.$$

One observes that the relations (3.1) and (3.2) are those of the unbound Kepler problem, which identifies the corresponding symmetry group as being SO(1, 3).

3.4 Four-vectors and multivectors in $\mathbb{H}(\mathbb{C})$

Let $x = (x_0 + i'\mathbf{x})$, $y = (y_0 + i'\mathbf{y})$ be two four-vectors and their conjugates x_c and y_c; one can define the exterior product

$$\begin{aligned} x \wedge y &= \frac{1}{2}(xy_c - yx_c) \\ &= \begin{bmatrix} 0, \\ (x_2y_3 - x_3y_2) + i'(x_1y_0 - x_0y_1), \\ (x_3y_1 - x_1y_3) + i'(x_2y_0 - x_0y_2), \\ (x_1y_2 - x_2y_1) + i'(x_3y_0 - x_0y_3) \end{bmatrix} \\ &= [0, \ \mathbf{x} \times \mathbf{y} + i'(y_0\mathbf{x} - x_0\mathbf{y})] \end{aligned}$$

with $x \wedge y = -y \wedge x$. The resulting quaternion is of the type $B = (\mathbf{a} + i'\mathbf{b})$ and is called a bivector; its real part, in the above example, gives the ordinary vector product but its nature differs from that of a four-vector. Under a Lorentz transformation (proper, orthochronous), a bivector transforms as

$$\begin{aligned} B' = x' \wedge y' &= \frac{1}{2}(x'y'_c - y'x'_c) \\ &= \frac{1}{2}[(axa^*_c)(a^*y_ca_c) - (aya^*_c)(a^*x_ca_c)] \\ &= aBa_c. \end{aligned}$$

The Lorentz transformation conserves the bivector type characterized by $B = -B_c$,

$$B'_c = aB_c a_c = -aBa_c = -B'.$$

Furthermore, $B^2 = \left[-(\mathbf{a})^2 + (\mathbf{b})^2 - 2i'\mathbf{a}\cdot\mathbf{b}\right]$ is a relativistic invariant

$$B'^2 = B'B' = aBa_c aBa_c = aB^2 a_c = B^2.$$

A trivector is a complex quaternion defined by

$$\begin{aligned} T = x \wedge B &= \frac{1}{2}\left(xB^* + Bx\right) \\ &= \left[-i'\mathbf{x}\cdot\mathbf{a} + x_0\mathbf{a} + \mathbf{x}\times\mathbf{b}\right] = \left(i't^0 + \mathbf{t}\right) \end{aligned}$$

and the product $B \wedge x$ by postulating $B \wedge x = x \wedge B$. Under a proper, orthochronous Lorentz transformation, one has

$$\begin{aligned} T' &= \frac{x'B'^* + B'x'}{2} \\ &= aTa_c^* \end{aligned}$$

that is, T transforms as a four-vector. Furthermore, the transformation conserves the trivector type $T_c^* = -T$,

$$\begin{aligned} T_c'^* &= aT_c^* a_c^* = a(-T)a_c^* \\ &= -T'; \end{aligned}$$

T yields the relativistic invariant

$$\begin{aligned} T'T'_c &= aTa_c^* a^* T_c a_c = TT_c \\ &= \left(t^0\right)^2 - \left(t^1\right)^2 - \left(t^2\right)^2 - \left(t^3\right)^2. \end{aligned}$$

The exterior product, defined above, is associative

$$(x \wedge y) \wedge z = x \wedge (y \wedge z).$$

Indeed,

$$(x \wedge y) \wedge z = z \wedge (x \wedge y) \tag{3.3}$$

$$= \frac{1}{2}\left[z\left(xy_c - yx_c\right)^* + \left(xy_c - yx_c\right)z\right] \tag{3.4}$$

$$= \frac{1}{2}\left[z\left(x_c y - y_c x\right) + \left(xy_c - yx_c\right)z\right], \tag{3.5}$$

$$x \wedge (y \wedge z) = \frac{1}{2}\left[x\left(yz_c - zy_c\right)^* + \left(yz_c - zy_c\right)x\right] \tag{3.6}$$

$$= \frac{1}{2}\left[x\left(y_c z - z_c y\right) + \left(yz_c - zy_c\right)z\right]. \tag{3.7}$$

Since $(z,x)y = y(z_c, x_c)$, one deduces the equality of the two equations (3.5), (3.7) and the associativity of the exterior product. A pseudoscalar is defined by the relation

$$P = x \wedge T = -\frac{1}{2}(xT_c + Tx^*)$$
$$= -\frac{1}{2}(xT_c + Tx_c)$$

where T is a trivector; by definition, one postulates $T \wedge x = -x \wedge T$. The pseudoscalar type is a pure imaginary $P = i's$ characterized by $P_c = P$ and is invariant under a Lorentz transformation

$$P' = -\frac{1}{2}(x'T'_c + T'x'_c) = aPa_c = P.$$

Examples. Consider the basis vectors $e_0 = 1$, $e_1 = i'i$, $e_2 = i'j$, $e_3 = i'k$; one obtains the following table

1	$i = e_2 \wedge e_3$	$j = e_3 \wedge e_1$	$k = e_1 \wedge e_2$
$i' = e_0 \wedge e_1 \wedge e_2 \wedge e_3$	$i'i = e_1 \wedge e_0$	$i'j = e_2 \wedge e_0$	$i'k = e_3 \wedge e_0$
$1 = e_0$	$i = e_0 \wedge e_2 \wedge e_3$	$j = e_0 \wedge e_3 \wedge e_1$	$k = e_0 \wedge e_1 \wedge e_2$
$i' = e_1 \wedge e_3 \wedge e_2$	$i'i = e_1$	$i'j = e_2$	$i'k = e_3$

One observes that distinct quantities occupy identical places in the $\mathbb{H}(\mathbb{C})$ algebra, a situation which one encounters also in the tridimensional vector calculus; this problem will be solved in the Clifford algebra $\mathbb{H} \otimes \mathbb{H}$. Besides the exterior products, one can define interior products

$$x \cdot y = (x,y) = \frac{1}{2}(xy_c + yx_c),$$
$$x \cdot B = \frac{1}{2}(xB^* - Bx) \in \text{four-vector},$$
$$x \cdot T = -\frac{1}{2}(xT_c - Tx^*) \in \text{bivector};$$

by definition, one supposes $B \cdot x = -x \cdot B$, $T \cdot x = x \cdot T$. More generally, one shall define the interior product between two multivectors A_p and B_q by [12, p. 14]

$$A_p \cdot B_q = (v_1 \wedge v_2 \cdots \wedge v_{p-1}) \cdot (v_p \cdot B_q).$$

Hence, one sees that $\mathbb{H}(\mathbb{C})$ already allows us to develop a few notions of a multivector calculus. These notions will be developed later on in the more satisfying framework of the Clifford algebra.

3.5 Relativistic kinematics via $\mathbb{H}(\mathbb{C})$

3.5.1 Special Lorentz transformation

Consider the reference frame at rest $K(O,x,y,z)$ and the reference frame $K'(O',x',y',z')$ moving along the Ox axis with the constant velocity $\mathbf{u}$ (Figure 3.1).

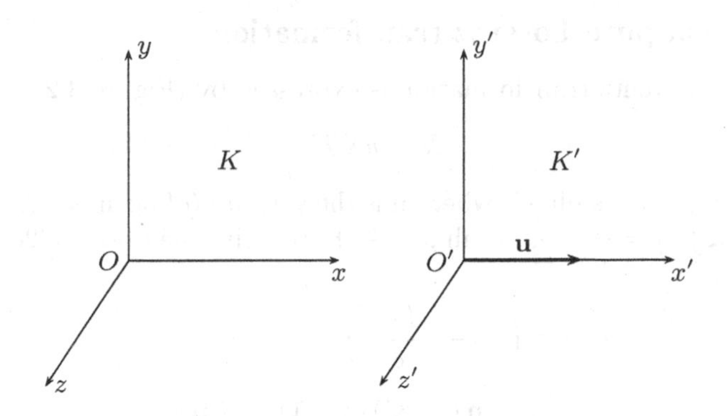

Figure 3.1: Special Lorentz transformation (pure): the axes remain parallel to themselves and the reference frame moves along the Ox axis.

The Lorentz transformation is expressed by

$$X = bX'b_c^* \tag{3.8}$$

with $X = (x^0 + i'\mathbf{x})$, $X' = (x'^0 + i'\mathbf{x}')$, $b = \left(\cosh\frac{\varphi}{2} + i'i\sinh\frac{\varphi}{2}\right)$, $bb_c = 1$, $\tanh\varphi = \frac{u}{c}$; write $\gamma = \cosh\varphi$, hence

$$\gamma^2 = 1 + \sinh^2\varphi = 1 + \frac{u^2}{c^2}\gamma^2,$$
$$\gamma = \frac{1}{\sqrt{1-\frac{u^2}{c^2}}}.$$

Explicitly, equation (3.8) reads

$$ct = \gamma\left(ct' + x'\frac{u}{c}\right),$$
$$x = \gamma\left(x' + ct'\frac{u}{c}\right),$$
$$y = y', \qquad z = z'.$$

The inverse transformation is given by $X' = b_c X b^*$ and yields

$$ct' = \gamma\left(ct - x\frac{u}{c}\right),$$
$$x' = \gamma\left(x - ct\frac{u}{c}\right),$$
$$y' = y, \qquad z' = z.$$

3.5.2 General pure Lorentz transformation

A general pure Lorentz transformation is expressed by (Figure 3.2)

$$X = bX'b_c^*$$

with $b = \left(\cosh\frac{\varphi}{2} + i'\frac{\mathbf{u}}{u}\sinh\frac{\varphi}{2}\right)$, where $\mathbf{u}$ is the velocity (of norm u), $X = (ct + i'\mathbf{x})$, $X' = (ct' + i'\mathbf{x}')$, $\gamma = \frac{1}{\sqrt{1-\frac{u^2}{c^2}}}$, $\tanh\varphi = \frac{u}{c}$. Explicitly, one obtains [26, p. 280]

$$ct = \gamma\left(ct' + \frac{\mathbf{x}'\cdot\mathbf{u}}{c}\right),$$
$$\mathbf{x} = \mathbf{x}' + \mathbf{n}\,(\mathbf{n}\cdot\mathbf{x}')\,(\gamma - 1) + ct'\mathbf{n}\gamma\frac{u}{c}.$$

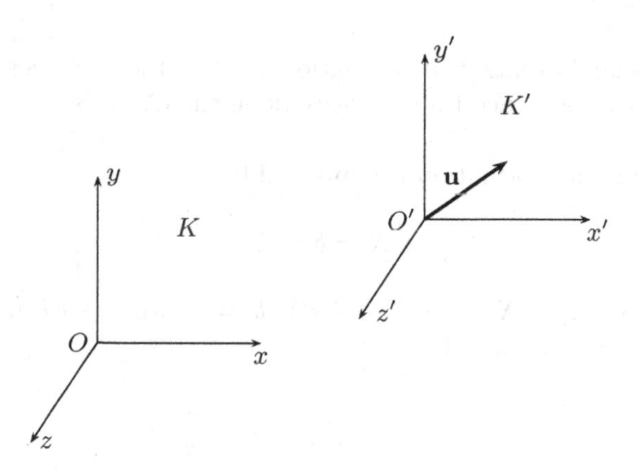

Figure 3.2: General pure Lorentz transformation: the axes remain parallel to themselves but the reference frame K' moves in an arbitrary direction.

3.5.3 Composition of velocities

Consider two reference frames K and K' with the special Lorentz transformation (Figure 3.1)

$$X = bX'b_c^*,$$

$b = \left(\cosh\frac{\varphi}{2} + i'i\sinh\frac{\varphi}{2}\right)$, $\tanh\varphi = \frac{u}{c}$, $\gamma = \cosh\varphi = \frac{1}{\sqrt{1-\frac{u^2}{c^2}}}$ (φ constant). The four-velocity V transforms as

$$V \equiv \frac{dX}{ds} = b\frac{dX'}{ds}b_c^*$$

with the relativistic invariant $ds = \sqrt{c^2dt^2 - dx^2 - dy^2 - dz^2}$; ds can also be expressed as

$$ds = \sqrt{c^2dt^2\left(1 - \frac{v^2}{c^2}\right)} = \frac{cdt}{\gamma} = \frac{cdt'}{\gamma'}.$$

The four-velocity satisfies the relation

$$VV_c = \frac{dX}{ds}\frac{dX_c}{ds} = 1$$

and can be written in the form

$$\begin{aligned} V &= \left(\cosh\theta + i'\frac{\mathbf{v}}{v}\sinh\theta\right) \\ &= \left(\gamma + i'\gamma\frac{\mathbf{v}}{c}\right) \end{aligned}$$

with $\tanh\theta = \frac{v}{c}$, $\gamma = \cosh\theta = \frac{1}{\sqrt{1-\frac{v^2}{c^2}}}$ and $\mathbf{v}$ the velocity (of norm v). Consider a particle moving along the Ox' axis with a velocity $\mathbf{v}'$ in the reference frame K', thus $V' = (\cosh\theta_1 + i'i\sinh\theta_1)$ with $\tanh\theta_1 = \frac{v'}{c}$. In the reference frame K, one has with $b = \left(\cosh\frac{\theta_2}{2} + i'i\sinh\frac{\theta_2}{2}\right)$, $\tanh\theta_2 = \frac{u}{c}$,

$$\begin{aligned} V &= bV'b_c^* \\ &= [\cosh(\theta_1 + \theta_2) + i'i\sinh(\theta_1 + \theta_2)] \\ &= (\cosh\theta + i'i\sinh\theta) \end{aligned}$$

hence, $\theta = \theta_1 + \theta_2$ and

$$\tanh\theta = \frac{v}{c} = \frac{\tanh\theta_1 + \tanh\theta_2}{1 + \tanh\theta_1\tanh\theta_2} = \frac{\frac{v'}{c} + \frac{u}{c}}{1 + \frac{v'u}{c^2}};$$

finally

$$v = \frac{v' + u}{1 + \frac{v'u}{c^2}}.$$

When $v', u \ll c$ one obtains the usual Galilean transformation. Furthermore, if $v' = c$, one has

$$v = \frac{c + u}{1 + \frac{u}{c}} = c,$$

the velocity of light thus appears as a limit speed.

3.6 Maxwell's equations

Let $A = (\frac{V}{c} + i'\mathbf{A})$ be the four-potential, D the relativistic four-nabla operator

$$D = \left(\frac{\partial}{c\partial t}, -i'\frac{\partial}{\partial x^1}, -i'\frac{\partial}{\partial x^2}, -i'\frac{\partial}{\partial x^3}\right)$$
$$= \left(\frac{\partial}{c\partial t} - i'\nabla\right)$$

and the conjugate operator

$$D_c = \left(\frac{\partial}{c\partial t} + i'\nabla\right).$$

Under a Lorentz transformation, D transforms as $x = (x^0 + i'\mathbf{x})$,

$$D' = aDa_c^*.$$

To verify this, it is sufficient to use the relation

$$\frac{\partial}{\partial x'^\mu} = \frac{\partial}{\partial x^\alpha}\frac{\partial x^\alpha}{\partial x'^\mu},$$

to develop $x' = axa_c^*$, $x = a_c x a^*$ and to compare the coefficients. Adopting the Lorentz gauge

$$(D, A) = \frac{1}{c^2}\frac{\partial V}{\partial t} + \operatorname{div}\mathbf{A} = 0$$

one obtains the electromagnetic field bivector

$$F = D_c A = (D, A) + (D \wedge A)$$
$$= (D \wedge A) = \left(-\mathbf{B} + \frac{i'\mathbf{E}}{c}\right).$$

Under a special Lorentz transformation, the bivector F transforms as

$$F' = bFb_c$$

with $b = \left(\cosh\frac{\varphi}{2} - i'i\sinh\frac{\varphi}{2}\right)$, $\tanh\varphi = \frac{v}{c}$, $\gamma = \cosh\varphi$, which yields the standard equations

$$B'_x = B_x, \qquad E'_x = E_x,$$

$$B'_y = \gamma\left(B_y + \frac{E_y v}{c^2}\right), \qquad E'_y = \gamma\left(E_y - vB_z\right),$$

$$B'_z = \gamma\left(B_z - \frac{E_y v}{c^2}\right), \qquad E'_z = \gamma\left(E_z + vB_y\right).$$

Furthermore, one has the relativistic invariant

$$FF_c = \left[\left(\mathbf{B}^2 - \frac{\mathbf{E}^2}{c^2}\right) + 2i'\frac{\mathbf{E}\cdot\mathbf{B}}{c}\right].$$

The exterior product

$$\begin{aligned} D \wedge F &= D \wedge D \wedge A = 0 \\ &= \left[-i' \operatorname{div} \mathbf{B}\,,\ -\frac{\partial \mathbf{B}}{c\partial t} - \frac{\operatorname{rot} \mathbf{E}}{c}\right] \end{aligned}$$

gives two of the Maxwell's equations

$$\operatorname{div} \mathbf{B} = 0, \qquad \operatorname{rot} \mathbf{E} = -\frac{\partial \mathbf{B}}{\partial t}.$$

Introducing the four-current density $C = (\rho c, i'\mathbf{j})$, the equation

$$\begin{aligned} D \cdot F &= \frac{DF^* - FD}{2} \\ &= \left[\frac{\operatorname{div}(\mathbf{E}}{c}\,,\ i'\left(\operatorname{rot} \mathbf{B} - \frac{1}{c^2}\frac{\partial \mathbf{E}}{\partial t}\right)\right] \\ &= \mu_0 C \end{aligned}$$

gives the two other Maxwell's equations (in vacuum)

$$\operatorname{div} \mathbf{E} = \frac{\rho}{\epsilon_0}, \qquad \operatorname{rot} \mathbf{H} = \mathbf{j} + \frac{\partial \mathbf{D}}{\partial t}$$

with $\epsilon_0 \mu_0 c^2 = 1$. Furthermore,

$$\begin{aligned} DF^* &= D \cdot F + D \wedge F \\ &= D(DA_c)^* = DD_c A \\ &= \Box A = \mu_0 C \end{aligned}$$

where

$$\Box = \frac{\partial^2}{c^2 \partial t^2} - \frac{\partial^2}{\partial\left(x^1\right)^2} - \frac{\partial^2}{\partial\left(x^2\right)^2} - \frac{\partial^2}{\partial\left(x^3\right)^2}$$

is the d'Alembertian. Hence, the equations

$$\begin{aligned} &\frac{\partial^2 V}{c^2 \partial t^2} - \triangle V = \frac{\rho}{\epsilon_0}, \\ &\frac{\partial^2 \mathbf{A}}{c^2 \partial t^2} - \triangle \mathbf{A} = \mu_0 \mathbf{j}. \end{aligned}$$

The entire set of Maxwell's equations (in vacuum) can therefore be written

$$\begin{aligned} DF^* &= \mu_0 C \\ &= \begin{bmatrix} \dfrac{\operatorname{div}\mathbf{E}}{c} - i'\operatorname{div}\mathbf{B} \\ +i'\left(\operatorname{rot}\mathbf{B} - \dfrac{1}{c^2}\dfrac{\partial\mathbf{E}}{\partial t}\right) - \left(\dfrac{\partial\mathbf{B}}{c\partial t} + \dfrac{\operatorname{rot}\mathbf{E}}{c}\right) \end{bmatrix} \\ &= \mu_0(\rho c + i'\mathbf{j}). \end{aligned}$$

The interior product

$$\begin{aligned} F\cdot C &= -C\cdot F \\ &= \left[\frac{\mathbf{j}\cdot\mathbf{E}}{c} + i'\rho\,(\mathbf{E} + \mathbf{v}\wedge\mathbf{B})\right] = f \end{aligned}$$

gives the volumic four-force (of Minkowski).

3.7 Group of conformal transformations

The group of conformal transformations is the group of transformations $x' = f(x)$ (x, x' being minquats) such that $dxdx_c = 0$ entails $dx'dx'_c = 0$. This group includes spacetime translations, Lorentz transformations, dilatations

$$\begin{aligned} x' &= x + d, \\ x' &= axa_c^*, \\ x' &= \lambda x \end{aligned}$$

($d \in$ minquat, $a \in \mathbb{H}(\mathbb{C})$, $aa_c = 1$, $\lambda \in \mathbb{R}$) and the transformations

$$x' = (1 + xa_c)^{-1}\,x \tag{3.9}$$

$$= x\,(1 + a_c x)^{-1} \tag{3.10}$$

$$= \frac{x + a(x,x)}{1 + 2(a,x) + (a,a)(x,x)} \tag{3.11}$$

with a respective number of parameters of $4, 6, 1, 4$ for a total of 15 parameters. The interest of this group in physics comes from the fact that Maxwell's equations (without sources) are covariant with respect to this transformation group. The transformations (3.11) can also be expressed in the form

$$\begin{aligned} (x')^{-1} &= x^{-1}\,(1 + xa_c)\,, \\ (x')^{-1} &= x^{-1} + a_c\,; \end{aligned}$$

the inverse transformation results from

$$\begin{aligned} x^{-1} &= (x')^{-1} - a_c \\ &= (x')^{-1}[1 - x'a_c] \end{aligned}$$

and thus

$$x = (1 - x'a_c)^{-1} x'.$$

The composition of two transformations gives

$$\begin{aligned} (x')^{-1} &= x^{-1} + a_c, \\ (x'')^{-1} &= (x')^{-1} + b_c \\ &= x^{-1} + a_c + b_c \\ &= x^{-1} + c_c \end{aligned}$$

with $c = a + b$ and thus belongs indeed to the group; if one permutes the two transformations, one obtains the same resulting transformation. As properties, one has

$$xa_cx' = x'a_cx, \tag{3.12}$$

$$|dx'|^2 = \frac{|x'|^2}{|x|^2}|dx|^2, \tag{3.13}$$

$$|x'|^2 = \frac{|x|^2}{|1 + xa_c|^2}, \tag{3.14}$$

$$dx' = (1 + xa_c)^{-1} dx\,(1 + a_cx)^{-1}. \tag{3.15}$$

Equation (3.12) results from

$$\begin{aligned} x' &= (1 + xa_c)^{-1} x = x(1 + a_cx)^{-1}, \\ (1 + xa_c)\, x' &= x = x'(1 + a_cx) \end{aligned}$$

which entails the relation. Furthermore,

$$\begin{aligned} x &= (1 + xa_c)x', \\ xx_c &= (1 + xa_c)\, x'x'_c\,(1 + ax_c), \end{aligned}$$

and thus

$$|x'|^2 = \frac{|x|^2}{|1 + xa_c|^2}.$$

Equation (3.13) which shows that the transformation is indeed a conformal transformation can be established as follows. Differentiating the relation $qq^{-1} = qq^{-1} = 1$, where q is a complex quaternion, one obtains

$$d\left(q^{-1}\right)q + q^{-1}dq = 0,$$
$$d\left(q^{-1}\right) = -q^{-1}dqq^{-1};$$

hence

$$d\left(x'^{-1}\right) = -\left(x'^{-1}\right)dx'\left(x'^{-1}\right)$$

or $d\left(x'^{-1}\right) = d\left(x^{-1}\right)$, consequently

$$\left(x'^{-1}\right)dx'\left(x'^{-1}\right) = \left(x^{-1}\right)dxx^{-1},$$
$$\frac{x'_c dx' x'_c}{\left(x'x'_c\right)^2} = \frac{x_c dx x_c}{\left(xx_c\right)^2}.$$

By multiplying with the conjugate equation

$$\frac{x' dx'_c x'}{\left(x'x'_c\right)^2} = \frac{x dx_c x}{\left(xx_c\right)^2}$$

one obtains equation (3.13)

$$\frac{dx' dx'_c}{\left(x'x'_c\right)^2} = \frac{dx dx_c}{\left(xx_c\right)^2}.$$

To obtain equation (3.15), one differentiates the equation

$$\begin{aligned}
x' &= x\left(1 + a_c x\right)^{-1}, \\
dx' &= dx\left(1 + a_c x\right)^{-1} - x\left(1 + a_c x\right)^{-1}\left[d\left(1 + a_c x\right)\right]\left(1 + a_c x\right)^{-1}, \\
&= \left[dx - \frac{x\left(1 + x_c a\right)a_c dx}{\left(1 + a_c x\right)\left(1 + x_c a\right)}\right]\left(1 + a_c x\right)^{-1}, \\
&= \frac{(1 + ax_c)dx\left(1 + a_c x\right)^{-1}}{\left(1 + a_c x\right)\left(1 + x_c a\right)}, \\
&= \left(1 + xa_c\right)^{-1}dx(1 + a_c x)^{-1}.
\end{aligned}$$

3.8 Exercises

E3-1 Express the matrices

$$e_1 = \begin{bmatrix} 1 & 0 \\ 0 & 0 \end{bmatrix}, \qquad e_2 = \begin{bmatrix} 0 & 0 \\ 0 & 1 \end{bmatrix}$$

using complex quaternions. Verify the relations $e_i^2 = e_i$, $e_1 + e_2 = 1$, $e_1 e_2 = e_2 e_1 = 0$. Take $\forall a \in \mathbb{H}(\mathbb{C})$, give the expression of the modules $u = ae_1$, $v = e_1 a$.

E3-2 Consider the complex quaternions

$$\begin{aligned} x &= 1 + i'i + (2 + i')j + k \\ y &= 2 + 3i'k; \end{aligned}$$

compute $x + y$, xy, yx, x^2, y^2, x^{-1}, y^{-1}, $y^{-1}x^{-1}$, $(xy)^{-1}$.

E3-3 Consider the general Lorentz transformation

$$\begin{aligned} X' &= aXa_c^*, \\ a &= \frac{1}{4}\left(3 + i'i\sqrt{5} - i'j\sqrt{15} + k3\sqrt{3}\right) \end{aligned}$$

with $aa_c = 1$. Determine the rotation r_i and the pure Lorentz transformation b_i such that $a = b_1 r_1 = r_2 b_2$ with

$$X' = b_1 r_1 X (b_1 r_1)_c^* = b_1 (r_1 X' r_{1c}) b_{1c}^*$$

or

$$X' = r_2 (b_2 X b_{2c}^*) r_{2c}.$$

E3-4 Consider the minquats

$$\begin{aligned} x &= 1 + i'(i + j), & y &= 2 + i'k, \\ z &= 3 + i'j, & w &= 3i'i + i'j + i'k. \end{aligned}$$

Compute $x \cdot y$, $B = x \wedge y$, $B' = z \wedge w$, $T = x \wedge (y \wedge z)$, $T' = (x \wedge y) \wedge z$, $B \cdot z$, $w \cdot T$, $B \wedge B'$, $B \cdot B'$, $B \cdot T$.

E3-5 Let $K(O, t, x, y, z)$ be a reference frame at rest and $K'(O', t', x', y', z')$ a reference frame moving along the Ox axis with a constant velocity $v = \frac{3\sqrt{5}}{7}c$. Write the Lorentz transformation $X = bX'b_c^*$ and give b. A particle is located at the point $X' = (0, i', i', 0)$ of K' and has a velocity $\frac{v'}{c} = \frac{1}{\sqrt{7}}$ in the direction $O'y'$; find its four-position and its four-velocity in the reference frame K. Let $\mathbf{E}'\left(E'_x, E'_y, E'_z\right)$ be the electric field in the reference frame K'; determine the electromagnetic field in K.

E3-6 Consider a square $ABCD$ of center 0 in the plane $z = 0$ and having its vertices located at the points $A(0, 1, -1, 0)$, $B(0, 1, 1, 0)$, $C(0, -1, 1, 0)$, $D(0, -1, -1, 0)$. Determine the transform of this square under the conformal transformation

$$x' = (1 + xa_c)^{-1} x \qquad \text{with } a = i'k.$$

Chapter 4

Clifford algebra

Clifford having demonstrated that the Clifford algebra is isomorphic to a tensor product of quaternion algebras or to a subalgebra thereof, this chapter develops within $\mathbb{H} \otimes \mathbb{H}$ the multivector calculus, multivectorial geometry and differential operators.

4.1 Clifford algebra

4.1.1 Definitions

The Clifford algebra C_n over $\mathbb{R}$ is an associative algebra having n generators $e_0, e_1, \ldots, e_{n-1}$ such that

$$e_i^2 = \pm 1, \qquad e_i e_j = -e_j e_i \qquad (i \neq j).$$

C^+ is the subalgebra constituted by products of an even number of e_i. Clifford algebras are directly related to the quaternion algebra via the following theorem ([13],[14]).

Theorem 4.1.1 (Clifford, 1878). *If $n = 2m$ (m integer), the Clifford algebra C_{2m} is the tensor product of m quaternion algebras. If $n = 2m-1$, the Clifford algebra C_{2m-1} is the tensor product of $m-1$ quaternion algebras and the algebra $(1, \omega)$ where ω is the product of the $2m$ generators ($\omega = e_0 e_1 \cdots e_{2m-1}$) of the algebra C_{2m} .*

The tensor product of the algebras A and B is defined as follows [8, p. 57]. Consider two algebras A and B with $x, y \in A$ and $u, v \in B$; the tensor product $A \otimes B$ is defined by the relation

$$(x \otimes u)\,(y \otimes v) = (xy) \otimes (uv)\,.$$

Example ($\mathbb{C} \otimes \mathbb{C}$). A general element of $\mathbb{C} \otimes \mathbb{C}$ is given by

$$\begin{aligned} A &= (a + i'b) \otimes (f + i'g) \\ &= (af)\,1 \otimes 1 + (bf)\,i' \otimes 1 + (ag)\,1 \otimes i' + (bg)\,i' \otimes i'; \end{aligned}$$

write $1 = 1 \otimes 1$, $i = i' \otimes 1$, $I = 1 \otimes i'$, $iI = Ii = i' \otimes i'$, one then has with $I^2 = -1$,

$$\begin{aligned} A &= (af + bfi) + I(ag + bgi) \\ &= (\alpha_1 + I\alpha_2). \end{aligned}$$

The general element of $\mathbb{C} \otimes \mathbb{C}$ can thus be expressed as a complex number having complex coefficients, the imaginary I commuting with i.

4.1.2 Clifford algebra $\mathbb{H} \otimes \mathbb{H}$ over $\mathbb{R}$

The general element of the tensor product of two quaternion algebras $\mathbb{H}$ is expressed by

$$\begin{aligned} A &= (a_0 + ia_1 + ja_2 + ka_3) \otimes (b_0 + ib_1 + jb_2 + kb_3) \\ &= \begin{bmatrix} a_0b_0 1 \otimes 1 + a_0b_1 1 \otimes i + a_0b_2 1 \otimes j + a_0b_3 1 \otimes k, \\ a_1b_0 i \otimes 1 + a_1b_1 i \otimes i + a_1b_2 i \otimes j + a_1b_3 i \otimes k, \\ a_2b_0 j \otimes 1 + a_2b_1 j \otimes i + a_2b_2 j \otimes j + a_2b_3 j \otimes k, \\ a_3b_0 k \otimes 1 + a_3b_1 k \otimes i + a_3b_2 k \otimes j + a_3b_3 k \otimes k \end{bmatrix}; \end{aligned}$$

let us write $i = i \otimes 1$, $j = j \otimes 1$, $k = k \otimes 1$, $I = 1 \otimes i$, $J = 1 \otimes j$, $K = 1 \otimes k$ with

$$\begin{aligned} &i^2 = j^2 = k^2 = ijk = -1, \\ &I^2 = J^2 = K^2 = IJK = -1. \end{aligned}$$

The general element A can thus be written

$$A = \alpha_0 + \alpha_1 I + \alpha_2 J + \alpha_3 K$$

where the coefficients $\alpha_i = d_i + ia_i + jb_i + kc_i$ are quaternions and where the lowercase i, j, k commute with the capital I, J, K ($iJ = Ji$, etc.). The element A is called a Clifford number and constitutes simply a quaternion having quaternions as coeffficients. Concisely, one can write

$$A = (\alpha_0; \boldsymbol{\alpha})$$

with $\boldsymbol{\alpha} = \alpha_1 I + \alpha_2 J + \alpha_3 K$, $\alpha_i \in \mathbb{H}$. The product of two Clifford numbers is given by

$$(\alpha_0; \boldsymbol{\alpha})(\beta_0; \boldsymbol{\beta}) = \begin{bmatrix} \alpha_0\beta_0 - \alpha_1\beta_1 - \alpha_2\beta_2 - \alpha_3\beta_3; \\ \alpha_0\beta_1 + \alpha_1\beta_0 + \alpha_2\beta_3 - \alpha_3\beta_2, \\ \alpha_0\beta_2 + \alpha_2\beta_0 + \alpha_3\beta_1 - \alpha_1\beta_3, \\ \alpha_0\beta_3 + \alpha_3\beta_0 + \alpha_1\beta_2 - \alpha_2\beta_1 \end{bmatrix}$$

and in a more compact notation

$$(\alpha_0; \boldsymbol{\alpha})(\beta_0; \boldsymbol{\beta}) = [\alpha_0\beta_0 - \boldsymbol{\alpha} \cdot \boldsymbol{\beta}; \alpha_0\boldsymbol{\beta} + \boldsymbol{\alpha}\beta_0 + \boldsymbol{\alpha} \times \boldsymbol{\beta}]$$

where $\boldsymbol{\alpha} \cdot \boldsymbol{\beta}$, $\boldsymbol{\alpha} \times \boldsymbol{\beta}$ are the ordinary scalar and vector products; the order of the terms has to be respected, the product of two quaternions being noncommutative. The generators of the Clifford algebra are

$$e_0 \equiv j, \qquad e_1 \equiv kI, \qquad e_2 \equiv kJ, \qquad e_3 \equiv kK$$

with $e_0^2 = -1$, $e_1^2 = e_2^2 = e_3^2 = 1$ and $e_i e_j = -e_j e_i$ $(i \neq j)$. A complete basis of the algebra is given in the following table.

1	$I = e_3 e_2$	$J = e_1 e_3$	$K = e_2 e_1$
$i = e_0 e_1 e_2 e_3$	$iI = e_0 e_1$	$iJ = e_0 e_2$	$iK = e_0 e_3$
$j = e_0$	$jI = e_0 e_3 e_2$	$jJ = e_0 e_1 e_3$	$jK = e_0 e_2 e_1$
$k = e_1 e_2 e_3$	$kI = e_1$	$kJ = e_2$	$kK = e_3$

A Clifford number can be written in the form [54]

$$A = (a + ib + jc + kd; \mathbf{m} + i\mathbf{n} + j\mathbf{r} + k\mathbf{s})$$

with $\mathbf{m} = m_1 I + m_2 J + m_3 K$ and similarly for $\mathbf{n}$, $\mathbf{r}$ and $\mathbf{s}$. The Clifford algebra contains scalars a, pseudoscalars ib, vectors (four-dimensional) $jc + k\mathbf{s}$, bivectors $\mathbf{m} + i\mathbf{n}$ and trivectors $kd + j\mathbf{r}$. The conjugate of A is defined by transforming I, J, K into $-I, -J, -K$ and j into $-j$, hence

$$A_c = (a + ib - jc + kd; -\mathbf{m} - i\mathbf{n} + j\mathbf{r} - k\mathbf{s})$$

with

$$(AB)_c = B_c A_c.$$

The dual of A is defined by

$$A^* = iA$$

and the commutator of two Clifford numbers by

$$[A, B] = \frac{1}{2}(AB - BA).$$

4.2 Multivector calculus within $\mathbb{H} \otimes \mathbb{H}$

4.2.1 Exterior and interior products with a vector

The product having been defined in the Clifford algebra, the interior and exterior products of two vectors $x = jx^0 + k\mathbf{x}$ $(\mathbf{x} = x^1 I + x^2 J + x^3 K)$, $y = jy^0 + k\mathbf{y}$ can be defined by the general formula [37]

$$xy = \lambda x \cdot y + \mu x \wedge y$$

where λ, μ are two nonzero coefficients. Adopting the choice $\lambda = \mu = -1$, one has

$$xy = -(x \cdot y + x \wedge y),$$
$$yx = -(y \cdot x + y \wedge x);$$

postulating a priori the relations $x \cdot y = y \cdot x$ and $x \wedge y = -y \wedge x$ one obtains

$$2x \cdot y = -(xy + yx),$$
$$2x \wedge y = -(xy - yx).$$

Explicitly, the formulas read

$$\begin{aligned} x \cdot y &= x^0y^0 - x^1y^1 - x^2y^2 - x^3y^3 \in S \quad \text{(scalar)}, \\ x \wedge y &= \begin{bmatrix} (x^2y^3 - x^3y^2)\, I + (x^3y^1 - x^1y^3)\, J + (x^1y^2 - x^2y^1)\, K \\ + (x^1y^0 - x^0y^1)\, iI + (x^2y^0 - x^0y^2)\, iJ + (x^3y^0 - x^0y^3)\, iK \end{bmatrix} \\ &= \left[\mathbf{x} \times \mathbf{y} + i\left(\mathbf{x}y^0 - x^0\mathbf{y}\right)\right] \in B \end{aligned}$$

with $x_c = -x$, $B_c = -B$ for a bivector B. The products of a vector with a multivector $A_p = v_1 \wedge v_2 \wedge \cdots \wedge v_p$ are then defined by

$$2x \cdot A_p = (-1)^p \left[xA_p - (-1)^p A_p x\right],$$
$$2x \wedge A_p = (-1)^p \left[xA_p + (-1)^p A_p x\right]$$

and

$$A_p \cdot x \equiv (-1)^p x \cdot A_p,$$
$$A_p \wedge x \equiv (-1)^p x \wedge A_p.$$

The above formulas yield for a trivector (with $x = jx^0 + k\mathbf{x}$, $B = \mathbf{a} + i\mathbf{b}$)

$$\begin{aligned} T = x \wedge B &= \frac{1}{2}(xB + Bx) \\ &= \left[(-\mathbf{a} \cdot \mathbf{x})\, k + j\left(x^0\mathbf{a} + \mathbf{x} \times \mathbf{b}\right)\right] \end{aligned}$$

with $B \wedge x = x \wedge B$ and $T_c = T$. One verifies that the exterior product is indeed associative

$$(x \wedge y) \wedge z = x \wedge (y \wedge z);$$

one has,

$$\begin{aligned} (x \wedge y) \wedge z &= z \wedge (x \wedge y) \\ &= -\frac{1}{2}\left[z(xy - yx) + (xy - yx)z\right], \end{aligned} \tag{4.1}$$

$$x \wedge (y \wedge z) = -\frac{1}{2}\left[x(yz - zy) + (yz - zy)x\right]. \tag{4.2}$$

Since $(z, x)y = y(z, x)$, i.e.,

$$(zx + xz)y = y(zx + xz),$$

one verifies the equality of the equations (4.1), (4.2) and thus the associativity. A pseudoscalar is defined by (with $T = kt^0 + j\mathbf{t}$, $x = jx^0 + k\mathbf{x}$)

$$\begin{aligned} P = x \wedge T &= -\frac{1}{2}(xT - Tx) \\ &= -\left(x^0 t^0 + \mathbf{x} \cdot \mathbf{t}\right) i \end{aligned}$$

with $T \wedge x \equiv -x \wedge T$ and $P_c = P$. As examples, let us express the standard basis in terms of the exterior product:

1	$I = e_2 \wedge e_3$	$J = e_3 \wedge e_1$	$K = e_1 \wedge e_2$
$i = e_0 \wedge e_1 \wedge e_2 \wedge e_3$	$iI = e_1 \wedge e_0$	$iJ = e_2 \wedge e_0$	$iK = e_3 \wedge e_0$
$j = e_0$	$jI = e_0 \wedge e_2 \wedge e_3$	$jJ = e_0 \wedge e_3 \wedge e_1$	$jK = e_0 \wedge e_1 \wedge e_2$
$k = e_1 \wedge e_3 \wedge e_2$	$kI = e_1$	$kJ = e_2$	$kK = e_3$

The interior products between a vector and a multivector are given by the formulas (with $x = jx^0 + k\mathbf{x}$, $B = \mathbf{a} + i\mathbf{b}$, $T = kt^0 + j\mathbf{t}$, $P = is$)

$$\begin{aligned} x \cdot B &= \frac{1}{2}(xB - Bx) \\ &= \left[(-\mathbf{b} \cdot \mathbf{x})\, j + k\left(-x^0 \mathbf{b} + \mathbf{x} \times \mathbf{a}\right)\right] \in V, \\ x \cdot T &= -\frac{1}{2}(xT + Tx) \\ &= \left[\left(x^0 \mathbf{t} + t^0 \mathbf{x}\right) + i\left(\mathbf{x} \times \mathbf{t}\right)\right] \in B, \\ x \cdot P &= \frac{1}{2}(xP - Px) \\ &= \left[-ksx^0 + j\left(s\mathbf{x}\right)\right] \in T \end{aligned}$$

with

$$\begin{aligned} B \cdot x &\equiv -x \cdot B, \\ T \cdot x &\equiv x \cdot T, \\ P \cdot x &\equiv -x \cdot P. \end{aligned}$$

4.2.2 Products of two multivectors

The products of two multivectors $A_p = v_1 \wedge v_2 \wedge \cdots \wedge v_p$ and $B_q = w_1 \wedge w_2 \wedge \cdots \wedge w_p$ are defined [12] for $p \leq q$, by

$$\begin{aligned} A_p \cdot B_q &\equiv (v_1 \wedge v_2 \wedge \cdots \wedge v_{p-1}) \cdot (v_p \cdot B_q), \\ A_p \wedge B_q &\equiv v_1 \wedge (v_2 \wedge \cdots \wedge v_p) \wedge B_q \end{aligned}$$

with

$$A_p \cdot B_q = (-1)^{p(q+1)} B_q \cdot A_p$$

which defines $B_q \cdot A_p$ for $q \geq p$. For products of multivectors one obtains, with S, P designating respectively the scalar and pseudoscalar parts of the multivector with $B = \mathbf{a} + i\mathbf{b}$, $B' = \mathbf{a}' + i\mathbf{b}'$, $T = ks_0 + j\mathbf{s}$, $T' = ks_0' + j\mathbf{s}'$, $P = iw$, $P' = iw'$,

$$\begin{aligned} B \cdot B' &= S\left[\frac{1}{2}(BB' + B'B)\right] \\ &= [-\mathbf{a} \cdot \mathbf{a}' + \mathbf{b} \cdot \mathbf{b}'] \in S, \end{aligned}$$

$$\begin{aligned} B \wedge B' &= P\left[\frac{1}{2}(BB' + B'B)\right] \\ &= i\,[-\mathbf{b} \cdot \mathbf{a}' - \mathbf{a} \cdot \mathbf{b}'] \in P, \end{aligned}$$

$$\begin{aligned} T \cdot T' &= -\frac{1}{2}(TT' + T'T) \\ &= s^0 s'^0 - s^1 s'^1 - s^2 s'^2 - s^3 s'^3 \in S, \end{aligned}$$

$$\begin{aligned} B \cdot P &= \frac{1}{2}(BP + PB) \\ &= (-s\mathbf{b} + is\mathbf{a}) \in B, \end{aligned}$$

$$\begin{aligned} T \cdot P &= \frac{1}{2}(TP - PT) \\ &= \left(jws^0 - kw\mathbf{s}\right) \in V, \end{aligned}$$

$$\begin{aligned} P \cdot P' &= \frac{1}{2}(PP' + P'P) \\ &= -ww' \in S. \end{aligned}$$

4.2.3 General formulas

Among general formulas, one has

$$x \cdot (y \wedge z) = (x \cdot y)z - (x \cdot z)y, \tag{4.3}$$

$$x \cdot (y \wedge z) + y \cdot (z \wedge x) + z \cdot (x \wedge y) = 0, \tag{4.4}$$

$$(x \wedge y) \cdot B = x \cdot (y \cdot B) = -y \cdot (x \cdot B), \tag{4.5}$$

$$(x \wedge y) \cdot (z \wedge w) = (x \cdot w)(y \cdot z) - (x \cdot z)(y \cdot w), \tag{4.6}$$

$$(B \cdot T) \cdot V = (B \wedge V) \cdot T, \tag{4.7}$$

the Jacobi identity

$$0 = [F, [G, H]] + [G, [H, F]] + [H, [F, G]]\,, \tag{4.8}$$

$$(x \cdot B_1) \cdot B_2 = B_1 \cdot (B_2 \cdot x) + x \cdot [B_1, B_2]\,, \tag{4.9}$$

with x, y, z, w being (four-)vectors, B, B_1, B_2 bivectors, T a trivector and F, G, H any Clifford numbers; the equations (4.4), (4.6), (4.9) are respectively consequences of equations (4.3), (4.5), (4.8). To establish equation (4.3), one just needs to write for any bivector B, $2x \cdot B = (xB - Bx)$, hence

$$4x \cdot (y \wedge z) = [-x(yz - zy) + (yz - zy)x]\,;$$

furthermore,

$$\begin{aligned} 4(x \cdot y)z &= -\left[(xy + yx)z + z(xy + yx)\right], \\ -4(x \cdot z)y &= \left[(xz + zx)y + y(xz + zx)\right]; \end{aligned}$$

adding the two last equations, one verifies equation (4.3). Equation (4.5) simply results from

$$\begin{aligned} (x \wedge y) \cdot B &\equiv x \cdot (y \cdot B) \\ &= -(y \wedge x) \cdot B \\ &= -y \cdot (x \cdot B); \end{aligned}$$

in particular,

$$(x \wedge y) \cdot (x \wedge y) = (x \cdot y)^2 - (x \cdot x)(y \cdot y).$$

To prove equation (4.7), one writes

$$\begin{aligned} 2B \cdot T &= (BT + TB), \\ 4(B \cdot T) \cdot V &= -\left[(BT + TB)V + V(BT + TB)\right], \\ 2B \wedge V &= 2V \wedge B = VB + BV, \\ 4(B \wedge V) \cdot T &= -\left[(VB + BV)T + T(VB + BV)\right]; \end{aligned}$$

or $B(TV - VT) = (TV - VT)B$ because $TV - VT$ is a pseudoscalar which commutes with B, hence the equation. The Jacobi identity results from

$$\begin{aligned} [A, BC] &= \frac{1}{2}\,(ABC - BCA) \\ &= \frac{1}{2}\,[(AB - BA)\,C + (BAC - BCA)] \\ &= [A, B]\,C + B\,[A, C]\,, \\ [A, CB] &= [A, C]\,B + C\,[A, B]\,, \\ [A, [B, C]] &= \frac{1}{2}\,\{[A, B]\,C + B\,[A, C] - [A, C]\,B - C\,[A, B]\} \\ &= [[A, B]\,, C] + [B, [A, C]] \end{aligned}$$

hence, the equation

$$[A,[B,C]]+[B,[C,A]]+[C,[A,B]]=0.$$

4.2.4 Classical vector calculus

The link with the classical vector calculus is obtained as follows. Let $x = k\mathbf{x}$, $y = k\mathbf{y}$, $z = k\mathbf{z}$, $w = k\mathbf{w}$ be four vectors without a temporal component, one obtains the relations

$$\begin{aligned}
x \cdot (y \wedge z) &= k\,[\mathbf{x} \times (\mathbf{y} \times \mathbf{z})] \in V,\\
x \wedge (y \wedge z) &= -k\,[\mathbf{x} \cdot (\mathbf{y} \times \mathbf{z})] \in T,\\
(x \wedge y) \cdot (z \wedge w) &= -(\mathbf{x} \times \mathbf{y}) \cdot (\mathbf{z} \times \mathbf{w}),\\
(x \wedge y) \cdot (x \wedge y) &= (x \cdot y)^2 - (x \cdot x)(y \cdot y),\\
&= (\mathbf{x} \cdot \mathbf{y})^2 - (\mathbf{x} \cdot \mathbf{x})\,(\mathbf{y} \cdot \mathbf{y}),\\
&= -(\mathbf{x} \times \mathbf{y}) \cdot (\mathbf{x} \times \mathbf{y}) \in S,\\
[x \wedge y, z \wedge w] &= (\mathbf{x} \times \mathbf{y}) \times (\mathbf{z} \times \mathbf{w}) \in B.
\end{aligned}$$

The entire classical vector calculus thus constitutes a particular case of the multivector calculus.

4.3 Multivector geometry

4.3.1 Analytic geometry

Straight line

In the space of four dimensions, the equation of a straight line parallel to the vector $u = ju_0 + k\mathbf{u}$ and going through the point $a = ja_0 + k\mathbf{a}$ is given by

$$(x-a) \wedge u = 0$$

with the immediate solution

$$\begin{aligned}
x - a &= \lambda u,\\
x &= \lambda u + a \qquad (\lambda \in \mathbb{R})
\end{aligned}$$

constituting the parametric equation of the straight line.

Plane

Let u, v be two linearly independent vectors and $B = u \wedge v$ the corresponding plane; the equation of a plane parallel to B and going through the point a is expressed by

$$(x-a) \wedge (u \wedge v) = 0$$

with the solution

$$\begin{aligned} x - a &= \lambda u + \mu v, \\ x &= \lambda u + \mu v + a \qquad (\lambda, \mu \in \mathbb{R}) \end{aligned}$$

giving the parametric equation of the plane parallel to B. A vector n is perpendicular to the plane $B = u \wedge v$ if n is perpendicular to u and v ($n \cdot u = n \cdot v = 0$); for any vector x one has

$$x \cdot (u \wedge v) = (x \cdot u)v - (x \cdot v)u$$

hence

$$n \cdot B = 0.$$

Furthermore, one remarks that the vector $x \cdot (u \wedge v)$ is perpendicular to x,

$$\begin{aligned} (x \cdot B) \cdot x &= (x \cdot u)(v \cdot x) - (x \cdot v)(u \cdot x) \\ &= 0. \end{aligned}$$

A plane $B_1 = x \wedge y$ is perpendicular to a plane $B_2 = u \wedge v$ if the vectors x, y are perpendicular to the vectors u, v ($x \cdot u = y \cdot u = x \cdot v = y \cdot v = 0$); from the general formula

$$(x \wedge y) \cdot (u \wedge v) = (x \cdot v)(y \cdot u) - (x \cdot u)(y \cdot v)$$

one obtains the orthogonality condition of two planes

$$B_1 \cdot B_2 = 0.$$

In particular, the dual of a plane $B^* = iB$ is perpendicular to that plane

$$B^* \cdot B = 0.$$

Hyperplane

A hyperplane is a subvector space of dimensions $p = n - 1 = 3$ (for $n = 4$). Let $T = u \wedge v \wedge w$ be a trivector (hyperplane), the equation of a hyperplane parallel to T and going through the point a is given by

$$(x - a) \wedge T = 0 \tag{4.10}$$

with the solution

$$\begin{aligned} x - a &= \lambda u + \mu v + \gamma w, \\ x &= \lambda u + \mu v + \gamma w + a \qquad (\lambda, \mu, \gamma \in \mathbb{R}) \end{aligned}$$

giving the parametric equation of the hyperplane. Explicitly, equation (4.10) reads (with $T = kt^0 + j\mathbf{t}$, $x = jx^0 + k\mathbf{x}$, $a = ja^0 + k\mathbf{a}$)

$$t^0 x^0 + t^1 x^1 + t^2 x^2 + t^3 x^3 - (a^0 t^0 + \mathbf{a} \cdot \mathbf{t}) = 0$$

which is indeed the equation of a hyperplane. A vector n is perpendicular to the hyperplane $T = u \wedge v \wedge w$ if n is perpendicular to u, v, w $(n \cdot u = n \cdot v = n \cdot w = 0)$. From the general formula

$$n \cdot (u \wedge v \wedge w) = (n \cdot u)(v \wedge w) + (n \cdot v)(w \wedge u) + (n \cdot w)(u \wedge w)$$

one deduces

$$n \cdot T = 0. \tag{4.11}$$

If n is the dual of T, $n = T^* = iT$, one finds (with $T = kt^0 + j\mathbf{t}$, $T^* = -jt^0 + k\mathbf{t}$)

$$T^* \cdot T = -\frac{1}{2}(T^*T + TT^*) = 0;$$

the dual of a hyperplane is thus perpendicular to that hyperplane. A plane $B = x \wedge y$ is perpendicular to the hyperplane $T = u \wedge v \wedge w$ if x, y are perpendicular to the hyperplane, hence

$$T \cdot B = B \cdot T = (x \wedge y) \cdot T \tag{4.12}$$
$$\equiv x \wedge (y \cdot T) = -y \wedge (x \cdot T) = 0. \tag{4.13}$$

4.3.2 Orthogonal projections

Orthogonal projection of a vector on a vector

Let $u = u_\| + u_\perp$ be the vector to project on the vector a with $u_\perp \cdot a = 0$, $u_\| \wedge a = 0$; since

$$ua = -u \cdot a - u \wedge a$$

one obtains

$$\begin{aligned} u_\| a &= -u_\| \cdot a = -u \cdot a, \\ u_\perp a &= -u_\perp \wedge a = -u \wedge a, \\ u_\| &= -(u \cdot a)a^{-1}, \\ u_\perp &= -(u \wedge a)a^{-1}. \end{aligned}$$

Orthogonal projection of a vector on a plane

Let $u = u_\| + u_\perp$ be the vector and let us represent the plane by a bivector $B = a \wedge b$ (with $u_\perp \cdot B = 0$, $u_\| \wedge B = 0$); since

$$\begin{aligned} uB &= u \cdot B + u \wedge B, \\ Bu &= B \cdot u + u \wedge B, \end{aligned}$$

one obtains

$$\begin{aligned} u_\| &= (u \cdot B)B^{-1} = B^{-1}(B \cdot u), \\ u_\perp &= (u \wedge B)B^{-1} = B^{-1}(u \wedge B). \end{aligned}$$

Orthogonal projection of a vector on a hyperplane

Let T be the hyperplane $u = u_\| + u_\perp$ with $u_\perp \cdot T = 0$, $u_\| \wedge T = 0$; one has

$$\begin{aligned} uT &= -u \cdot T - u \wedge T, \\ Tu &= -u \cdot T + u \wedge T, \end{aligned}$$

hence

$$\begin{aligned} u_\| T &= -u \cdot T, & Tu_\| &= -u \cdot T, \\ u_\perp T &= -u \wedge T, & Tu_\perp &= u \wedge T, \end{aligned}$$

and finally

$$u_\| = -(u \cdot T)T^{-1} = -T^{-1}(u \cdot T), \tag{4.14}$$

$$u_\perp = -(u \wedge T)T^{-1} = T^{-1}(u \wedge T). \tag{4.15}$$

Orthogonal projection of a bivector on a plane

Let $B_1 = B_{1\|} + B_{1\perp}$ be the bivector and $B_2 = a \wedge b$ the plane with $B_{1\perp} \cdot B_2 = 0$, $B_{1\|} \wedge B_2 = 0$ and $\left[B_{1\|}, B_2\right] = 0$; using the formula

$$B_1 B_2 = B_1 \cdot B_2 + B_1 \wedge B_2 + [B_1, B_2],$$

one obtains

$$\begin{aligned} B_{1\|} &= (B_1 \cdot B_2) B_2^{-1}, \\ B_{1\perp} &= \{(B_1 \wedge B_2) + [B_1, B_2]\} B_2^{-1}. \end{aligned}$$

Example. Take $B_1 = x \wedge y$ (with $x = jx^0 + k\mathbf{x}$, $y = jy^0 + k\mathbf{y}$) and let us project on the plane x^1x^2, i.e., the plane $B_2 = K$ ($B_2^{-1} = -K$); one finds

$$\begin{aligned} B_{1\|} &= (B_1 \cdot B_2) B_2^{-1} \\ &= (-x^2y^1 + x^1y^2)K = x_\| \wedge y_\| \end{aligned}$$

(with $x_\| = x^1 kI + x^2 kJ$, $y_\| = y^1 kI + y^2 kJ$),

$$\begin{aligned} B_{1\perp} &= \{(B_1 \wedge B_2) + [B_1, B_2]\} B^{-1} \\ &= (-x_3y_2 + x_2y_3)I + (x_3y_1 - x_1y_3)J \\ &\quad + (x_1y_0 - x_0y_1)iI + (x_2y_0 - x_0y_2)iJ + (x_3y_0 - x_0y_3)iK \\ &= B_1 - B_{1\|}. \end{aligned}$$

Orthogonal projection of a bivector on a vector

Let $B = B_{\|} + B_{\perp}$ be the bivector and a the vector (with $B_{1\|} \wedge a = 0$, $B_{\perp} \cdot a = 0$); since

$$aB = a \cdot B + a \wedge B,$$
$$Ba = B \cdot a + a \wedge B,$$

one obtains

$$B_{\|} = a^{-1}(a \cdot B) = (B \cdot a)a^{-1},$$
$$B_{\perp} = a^{-1}(a \wedge B) = (a \wedge B)a^{-1}.$$

Orthogonal projection of a bivector on a hyperplane

Let $B = B_{\|} + B_{\perp}$ be the bivector and T the hyperplane with $B_{\perp} \cdot T = 0$ and $[B_{\|}, T] = 0$ $(B_{\|} \in T)$. From the equation

$$BT = B \cdot T + [B, T]$$

one obtains

$$B_{\|} = (B \cdot T)T^{-1},$$
$$B_{\perp} = [B, T]\, T^{-1}.$$

Example. Let $F = -\mathbf{B} + i\frac{\mathbf{E}}{c}$ be the electromagnetic bivector, its orthogonal projection on the hyperplane $T = k = e_1e_2e_3$ $(T^{-1} = -k)$ yields

$$F_{\|} = -(F \cdot k)k = -\mathbf{B},$$
$$F_{\perp} = -[F, k]\, k = \frac{\mathbf{E}}{c}.$$

Orthogonal projection of a hyperplane T_1 on the hyperplane T_2

Let us write $T_1 = T_{1\|} + T_{1\perp}$ with $T_1 = x \wedge y \wedge z$, $T_2 = u \wedge v \wedge w$; from the relation

$$T_1 \cdot T_2 \equiv (x \wedge y) \cdot [z \cdot (u \wedge v \wedge w)]$$

and the equation (4.11), one deduces

$$T_{1\perp} \cdot T_2 = 0;$$

furthermore $\left[T_{1\|}, T_2\right] = 0$ $(T_{1\|} \in T_2)$. Since

$$T_1T_2 = -T_1 \cdot T_2 + [T_1, T_2]$$

it follows that

$$T_{1\|} = -(T_1 \cdot T_2)T_2^{-1},$$
$$T_{1\perp} = [T_1, T_2]\, T_2^{-1}.$$

4.4 Differential operators

4.4.1 Definitions

Consider the relativistic four-nabla operator, ∇

$$\nabla = \left(j\frac{\partial}{\partial x^0} - k\boldsymbol{\nabla}\right)$$
$$= \partial_\alpha e^\alpha = \partial^\alpha e_\alpha$$

with

$$\boldsymbol{\nabla} = I\frac{\partial}{\partial x^1} + J\frac{\partial}{\partial x^2} + K\frac{\partial}{\partial x^3},$$

$\partial_\alpha = \frac{\partial}{\partial x^\alpha}$ and e^α the reciprocal basis defined by

$$e^\alpha \cdot e_\beta = \delta^\alpha_\beta$$

($e^0 = e_0$, $e^1 = -e_1$, $e^2 = -e_2$, $e^3 = -e_3$). One can define the operators four-gradient

$$\nabla\varphi = j\frac{\partial}{\partial x^0} - k\,\mathbf{grad}\,\varphi \qquad (\varphi \in S)$$

the four-divergence of a vector $A = jA^0 + k\mathbf{A}$

$$\nabla \cdot A = \frac{\partial A^0}{\partial x^0} + \frac{\partial A^1}{\partial x^1} + \frac{\partial A^2}{\partial x^2} + \frac{\partial A^3}{\partial x^3} \in S$$

the four-curl

$$\nabla \wedge A = -\,\mathbf{rot}\,\mathbf{A} - i\left(\frac{\partial \mathbf{A}}{\partial x^0} + \mathbf{grad}\,A^0\right) \in B;$$

acting on a bivector $B = \mathbf{a} + i\mathbf{b}$, one can define the operators

$$\nabla \cdot B = j\,\mathrm{div}\,\mathbf{b} - k\left(\frac{\partial \mathbf{b}}{\partial x^0} + \mathbf{rot}\,\mathbf{a}\right) \in V$$

$$\nabla \wedge B = k\,\mathrm{div}\,\mathbf{a} + j\left(\frac{\partial \mathbf{a}}{\partial x^0} - \mathbf{rot}\,\mathbf{b}\right) \in T.$$

4.4.2 Infinitesimal elements of curves, surfaces and hypersurfaces

A curve in the four-dimensional space is defined by the parametric equations

$$OM(\alpha) = jx^0(\alpha) + k\mathbf{x}(\alpha)$$

defining the tangent vector at the point M,

$$dOM = \frac{\partial OM}{\partial \alpha} d\alpha$$

where α is the parameter. For a surface, the parametric equations are

$$OM(\alpha, \beta) = jx^0(\alpha, \beta) + k\mathbf{x}(\alpha, \beta)$$

and the tangent plane to the surface at the point M is defined by the bivector

$$\begin{aligned} dS &= \left(\frac{\partial OM}{\partial \alpha} \wedge \frac{\partial OM}{\partial \beta} \right) d\alpha d\beta \\ &= \left\{ \begin{array}{l} \left(\frac{\partial x^2}{\partial \alpha} \frac{\partial x^3}{\partial \beta} - \frac{\partial x^3}{\partial \alpha} \frac{\partial x^2}{\partial \beta} \right) I + \left(\frac{\partial x^3}{\partial \alpha} \frac{\partial x^1}{\partial \beta} - \frac{\partial x^1}{\partial \alpha} \frac{\partial x^3}{\partial \beta} \right) J \\ + \left(\frac{\partial x^1}{\partial \alpha} \frac{\partial x^2}{\partial \beta} - \frac{\partial x^2}{\partial \alpha} \frac{\partial x^1}{\partial \beta} \right) K + \left(\frac{\partial x^1}{\partial \alpha} \frac{\partial x^0}{\partial \beta} - \frac{\partial x^0}{\partial \alpha} \frac{\partial x^1}{\partial \beta} \right) iI \\ + \left(\frac{\partial x^2}{\partial \alpha} \frac{\partial x^0}{\partial \beta} - \frac{\partial x^0}{\partial \alpha} \frac{\partial x^2}{\partial \beta} \right) iJ + \left(\frac{\partial x^3}{\partial \alpha} \frac{\partial x^0}{\partial \beta} - \frac{\partial x^0}{\partial \alpha} \frac{\partial x^3}{\partial \beta} \right) iK \end{array} \right\} d\alpha d\beta. \end{aligned}$$

In abridged notation, one can write

$$\begin{aligned} dS = {} & dx^2 dx^3 I + dx^3 dx^1 J + dx^1 dx^2 K \\ & + dx^1 dx^0 iI + dx^2 dx^0 iJ + dx^3 dx^0 iK \end{aligned}$$

with

$$\begin{aligned} dx^2 dx^3 &= \frac{D(x^2, x^3)}{D(\alpha, \beta)} d\alpha d\beta \\ &= \begin{vmatrix} \frac{\partial x^2}{\partial \alpha} & \frac{\partial x^2}{\partial \beta} \\ \frac{\partial x^3}{\partial \alpha} & \frac{\partial x^3}{\partial \beta} \end{vmatrix} d\alpha d\beta, \text{ etc.} \end{aligned}$$

where $dx^2 dx^3$ is an undissociable symbol [3, p. 446]. For a hypersurface defined by

$$OM(\alpha, \beta, \gamma) = jx^0(\alpha, \beta, \gamma) + k\mathbf{x}(\alpha, \beta, \gamma)$$

the tangent hyperplane to the surface at the point M is given by the trivector

$$\begin{aligned} dT &= \left(\frac{\partial OM}{\partial \alpha} \wedge \frac{\partial OM}{\partial \beta} \wedge \frac{\partial OM}{\partial \gamma} \right) d\alpha d\beta d\gamma \\ &= \left\{ \begin{array}{l} \left(\begin{array}{l} \frac{\partial x^3}{\partial \alpha} \frac{\partial x^2}{\partial \beta} \frac{\partial x^1}{\partial \gamma} - \frac{\partial x^2}{\partial \alpha} \frac{\partial x^3}{\partial \beta} \frac{\partial x^1}{\partial \gamma} - \frac{\partial x^3}{\partial \alpha} \frac{\partial x^1}{\partial \beta} \frac{\partial x^2}{\partial \gamma} \\ + \frac{\partial x^1}{\partial \alpha} \frac{\partial x^3}{\partial \beta} \frac{\partial x^2}{\partial \gamma} + \frac{\partial x^2}{\partial \alpha} \frac{\partial x^1}{\partial \beta} \frac{\partial x^3}{\partial \gamma} - \frac{\partial x^1}{\partial \alpha} \frac{\partial x^2}{\partial \beta} \frac{\partial x^3}{\partial \gamma} \end{array} \right) k \\ + \left(\begin{array}{l} - \frac{\partial x^3}{\partial \alpha} \frac{\partial x^2}{\partial \beta} \frac{\partial x^0}{\partial \gamma} + \frac{\partial x^2}{\partial \alpha} \frac{\partial x^3}{\partial \beta} \frac{\partial x^0}{\partial \gamma} + \frac{\partial x^3}{\partial \alpha} \frac{\partial x^0}{\partial \beta} \frac{\partial x^2}{\partial \gamma} \\ - \frac{\partial x^0}{\partial \alpha} \frac{\partial x^3}{\partial \beta} \frac{\partial x^2}{\partial \gamma} - \frac{\partial x^2}{\partial \alpha} \frac{\partial x^0}{\partial \beta} \frac{\partial x^3}{\partial \gamma} + \frac{\partial x^0}{\partial \alpha} \frac{\partial x^2}{\partial \beta} \frac{\partial x^3}{\partial \gamma} \end{array} \right) jI \\ + \left(\begin{array}{l} \frac{\partial x^3}{\partial \alpha} \frac{\partial x^1}{\partial \beta} \frac{\partial x^0}{\partial \gamma} - \frac{\partial x^1}{\partial \alpha} \frac{\partial x^3}{\partial \beta} \frac{\partial x^0}{\partial \gamma} - \frac{\partial x^3}{\partial \alpha} \frac{\partial x^0}{\partial \beta} \frac{\partial x^1}{\partial \gamma} \\ + \frac{\partial x^0}{\partial \alpha} \frac{\partial x^3}{\partial \beta} \frac{\partial x^1}{\partial \gamma} + \frac{\partial x^1}{\partial \alpha} \frac{\partial x^0}{\partial \beta} \frac{\partial x^3}{\partial \gamma} - \frac{\partial x^0}{\partial \alpha} \frac{\partial x^1}{\partial \beta} \frac{\partial x^3}{\partial \gamma} \end{array} \right) jJ \\ + \left(\begin{array}{l} - \frac{\partial x^2}{\partial \alpha} \frac{\partial x^1}{\partial \beta} \frac{\partial x^0}{\partial \gamma} + \frac{\partial x^1}{\partial \alpha} \frac{\partial x^2}{\partial \beta} \frac{\partial x^0}{\partial \gamma} + \frac{\partial x^2}{\partial \alpha} \frac{\partial x^0}{\partial \beta} \frac{\partial x^1}{\partial \gamma} \\ - \frac{\partial x^0}{\partial \alpha} \frac{\partial x^2}{\partial \beta} \frac{\partial x^1}{\partial \gamma} - \frac{\partial x^1}{\partial \alpha} \frac{\partial x^0}{\partial \beta} \frac{\partial x^2}{\partial \gamma} + \frac{\partial x^0}{\partial \alpha} \frac{\partial x^1}{\partial \beta} \frac{\partial x^2}{\partial \gamma} \end{array} \right) jK \end{array} \right\} d\alpha d\beta d\gamma. \end{aligned}$$

In short, one can write

$$dT = kdx^1dx^3dx^2 \tag{4.16}$$

$$+ jIdx^2dx^3dx^0 + jJdx^3dx^1dx^0 + jKdx^1dx^2dx^0 \tag{4.17}$$

with the symbol

$$dx^1dx^3dx^2 = \frac{D(x^1,x^3,x^2)}{D(\alpha,\beta,\gamma)}d\alpha d\beta d\gamma$$

$$= \begin{vmatrix} \frac{\partial x^1}{\partial \alpha} & \frac{\partial x^1}{\partial \beta} & \frac{\partial x^1}{\partial \gamma} \\ \frac{\partial x^3}{\partial \alpha} & \frac{\partial x^3}{\partial \beta} & \frac{\partial x^3}{\partial \gamma} \\ \frac{\partial x^2}{\partial \alpha} & \frac{\partial x^2}{\partial \beta} & \frac{\partial x^2}{\partial \gamma} \end{vmatrix} d\alpha d\beta d\gamma, \text{ etc.}$$

4.4.3 General theorems

Generalized Stokes theorem

This theorem can be written

$$\oint A \cdot dl = -\iint (\nabla \wedge A) \cdot dS \tag{4.18}$$

with $dl = jdx^0 + k\mathbf{dx}$, $A = jA^0 + k\mathbf{A}$, $dS = \mathbf{dS}_1 + i\mathbf{dS}_2$ and

$$\mathbf{dS}_1 = dx^2dx^3I + dx^3dx^1J + dx^1dx^2K,$$
$$\mathbf{dS}_2 = dx^1dx^0I + dx^2dx^0J + dx^3dx^0K.$$

Explicitly, equation (4.18) is expressed in classical vector notation as

$$\oint A^0dx^0 - \mathbf{A}\cdot\mathbf{dx} = \iint \left[-\mathbf{rot}\,\mathbf{A}\cdot\mathbf{dS}_1 + \left(\nabla A^0 + \frac{\partial \mathbf{A}}{\partial x^0}\right)\cdot\mathbf{dS}_2\right].$$

If $dx^0 = 0$, one has $\mathbf{dS}_2 = 0$ and the formula reduces to the standard Stokes theorem. As to the orientation of the curve C, it results from that of the surface S [10, p. 39]. One takes two linearly independent four-vectors $a, b \in S$ and one chooses the order (a, b) as being the positive orientation ($dS = \alpha a \wedge b$, $\alpha > 0$). On the curve C one chooses a vector f exterior to S and a vector g tangent to the curve such that (f, g) is ordered positively; the curve is then oriented along g.

Generalized Gauss theorem

The theorem is expressed as

$$\iiint A \wedge dT = \iiiint (\nabla \cdot A)\, d\tau \tag{4.19}$$

with $d\tau = idx^0dx^1dx^2dx^3$, $dT = kdx^1dx^3dx^2 + j\mathbf{t}$ and

$$\mathbf{t} = dx^2dx^3dx^0I + dx^3dx^1dx^0J + dx^1dx^2dx^0K,$$

the hypersurface being closed. Explicitly, equation (4.19) reads

$$\iiint \left(-A^0dx^1dx^3dx^2 + \mathbf{A}\cdot\mathbf{t}\right) i = \iiiint \left(\frac{\partial A^0}{\partial x^0} + \operatorname{div}\mathbf{A}\right) idx^0dx^1dx^2dx^3.$$

The orientation of the hypersurface results from that of the four-volume. Let a, b, c, d be four linearly independent vectors of the four-volume $a \wedge b \wedge c \wedge d = \alpha i$; if $\alpha > 0$, the orientation of the four-volume is positive. On a point of the hypersurface, one chooses a vector m outside of the four-volume and three vectors p, q, r on the hypersurface; if $m \wedge p \wedge q \wedge r$ has a positive sign, the orientation of the hypersurface is positive.

Other formulas

On a closed surface, one has

$$\iint F \cdot dS = \iiint (\nabla \wedge F) \cdot dT, \tag{4.20}$$

$$\iint F \wedge dS = -\iiint (\nabla \cdot F) \wedge dT, \tag{4.21}$$

with the bivector $F = \mathbf{f} + i\mathbf{g}$, $dS = \mathbf{dS}_1 + i\mathbf{dS}_2$, $dT = kdx^1dx^3dx^2 + j\mathbf{t}$; explicitly, formula (4.20) reads

$$\iint (-\mathbf{f}\cdot\mathbf{dS}_1 + \mathbf{g}\cdot\mathbf{dS}_2) = \iiint \left[\operatorname{div}\mathbf{f}dx^1dx^3dx^2 - \left(\frac{\partial\mathbf{f}}{\partial x^0} - \mathbf{rot}\,\mathbf{g}\right)\cdot\mathbf{t}\right];$$

if $dx^0 = 0$, $\mathbf{dS}_2 = 0$, $\mathbf{t} = 0$, one obtains the standard Gauss theorem. Relation (4.21) gives

$$\iint i\,(-\mathbf{g}\cdot\mathbf{dS}_1 - \mathbf{f}\cdot\mathbf{dS}_2) = \iiint i\left[\operatorname{div}\mathbf{g}dx^1dx^3dx^2 - \left(\frac{\partial\mathbf{g}}{\partial x^0} + \mathbf{rot}\,\mathbf{f}\right)\cdot\mathbf{t}\right].$$

The orientation of dS proceeds from that of dT. One chooses three linearly independent vectors a, b, c of T and one defines the orientation $a \wedge b \wedge c$ as being positive. On a point of the surface, one considers a vector m exterior to the trivolume T, and two vectors p, q on the surface; if $m \wedge p \wedge q$ has the orientation of $a \wedge b \wedge c$, the orientation of the surface is positive.

4.5 Exercises

E4-1 Consider the Clifford numbers A and B:

$$A = I + 2J - iK,$$
$$B = j + kI + 2kK.$$

Determine A_c, B_c, $A+B$, $A-B$, AA_c, BB_c, A^{-1}, B^{-1}, AB, BA, $(AB)_c$, $(BA)_c$, $(AB)(AB)_c$, $(BA)(BA)_c$, $(AB)^{-1}$, $(BA)^{-1}$, $[A,B]$.

E4-2 Consider the four-vectors

$$x = \jmath + kI + kJ, \qquad y = 2\jmath + kK,$$
$$z = 3\jmath + kJ, \qquad w = 3kI + kJ + kK.$$

Determine $x \cdot y$, $B = x \wedge y$, $B' = z \wedge w$, $T = x \wedge (y \wedge z)$, $T' = (x \wedge y) \wedge z$, $B \cdot z$, $w \cdot T$, $B \cdot B'$, $B \wedge B'$, $B \cdot T$.

E4-3 Take an orthonormal reference frame with the components of the four-vectors

$$x = (0, 1, 2, -1), \qquad y = (0, 3, 1, 1),$$
$$z = (0, 1, 2, 1), \qquad w = (0, 2, 1, 5).$$

Determine within the Clifford algebra $\mathbb{H} \otimes \mathbb{H}$ the surfaces $S_1 = x \wedge y$, $S_2 = z \wedge w$, the trivector $x \wedge y \wedge z$ and the four-volume $x \wedge y \wedge z \wedge w$. Give the orthogonal projection of the vector w on S_1 and the orthogonal projection of S_1 on S_2.

Chapter 5

Symmetry groups

This chapter formulates the Lorentz group and the group of conformal transformations within the Clifford algebra $\mathbb{H} \otimes \mathbb{H}$ over $\mathbb{R}$. In complexifying this algebra, one obtains the Dirac algebra $\mathbb{H} \otimes \mathbb{H}$ over $\mathbb{C}$, isomorphic to the subalgebra C^+ of $\mathbb{H} \otimes \mathbb{H} \otimes \mathbb{H}$ over $\mathbb{R}$. Dirac's equation, the unitary group SU(4) and the symplectic unitary group $\mathrm{USp}(2, \mathbb{H})$ are treated as applications of $\mathbb{H} \otimes \mathbb{H}$ over $\mathbb{C}$.

5.1 Pseudo-orthogonal groups $\mathrm{O}(1,3)$ and $\mathrm{SO}(1,3)$

5.1.1 Metric

Consider two vectors $x = jx^0 + k\mathbf{x}$, $y = jy^0 + k\mathbf{y}$ with $\mathbf{x} = x^1 I + x^2 J + x^3 K$ and the interior product, with $x_c = -x$;

$$\begin{aligned} x \cdot x &= x x_c = -x^2, \\ (x+y)\cdot(x+y) &= (x+y)(x+y)_c \\ &= x x_c + y y_c + 2x \cdot y, \\ x \cdot y &= \frac{(x y_c + y x_c)}{2} = -\frac{(xy + yx)}{2} \\ &= x^0 y^0 - x^1 y^1 - x^2 y^2 - x^3 y^3. \end{aligned}$$

A vector x is isotropic if $x x_c = 0$, timelike if $x x_c > 0$ and spacelike if $x x_c < 0$.

5.1.2 Symmetry with respect to a hyperplane

Let $a = ja^0 + k\mathbf{a}$ be a vector, the hyperplane perpendicular to a is given by the dual of a,

$$\begin{aligned} T &= a^* = ia \\ &= ka^0 - j\mathbf{a} \end{aligned}$$

with $T^{-1} = \frac{-ia}{aa_c}$ (i commutes with all elements of C^+ and anticommutes with those of C^-). Let us suppose that T goes through the origin and let $x = jx^0 + k\mathbf{x}$ be a vector. The orthogonal projections of x on T are given by the relations (4.14), (4.15)

$$x_{\|} = -T^{-1}(x \cdot T),$$
$$x_{\perp} = T^{-1}(x \wedge T),$$

hence

$$x_{\perp} = -T^{-1}\frac{(xT - Tx)}{2} = \frac{x}{2} + \frac{iaxia}{2aa_c} = \frac{x}{2} - \frac{axa}{2aa_c},$$
$$x_{\|} = \frac{x}{2} + \frac{axa}{2aa_c}$$

with $x = x_{\|} + x_{\perp}$.

Definition 5.1.1. The symmetric of x with respect to a hyperplane is obtained by drawing the perpendicular to the hyperplane and by extending this perpendicular by an equal length [11].

Let x' be the symmetric of x with respect to the hyperplane T ; one has

$$x - x' = 2x_{\perp}$$

hence

$$x' = \frac{axa}{aa_c}.$$

More simply, one can write that $x' - x$ is perpendicular to the hyperplane T (and thus parallel to a) and $\frac{x+x'}{2}$ is parallel to the hyperplane. One obtains

$$x' = x + \lambda a,$$
$$a \cdot \left(\frac{x' + x}{2}\right) = 0,$$

hence

$$a \cdot \left(x + \frac{\lambda a}{2}\right) = 0 \Longrightarrow \lambda = -\frac{2(a \cdot x)}{a \cdot a},$$
$$x' = x - \frac{2(a \cdot x)a}{a \cdot a}$$
$$= x + \frac{(ax + xa)a}{aa_c} = \frac{axa}{aa_c} = -\frac{ax_c a}{aa_c};$$

finally (with $a^2 = -aa_c$), one finds again the above expression of the symmetric of x with respect to the hyperplane. One shall distinguish the time symmetries

(with $aa_c = 1$)

$$\begin{aligned} x' &= axa \\ &= -ax_c a, \end{aligned}$$

from the space symmetries (with $aa_c = -1$)

$$\begin{aligned} x' &= -axa \\ &= ax_c a. \end{aligned}$$

5.1.3 Pseudo-orthogonal groups O(1,3) and SO(1,3)

Definition 5.1.2. The pseudo-orthogonal group $\mathrm{O}(1,3)$ is the group of linear operators which leave invariant the quadratic form

$$x \cdot y = x^0y^0 - x^1y^1 - x^2y^2 - x^3y^3$$

with $x = jx^0 + k\mathbf{x}$.

Theorem 5.1.3. *Every rotation of* $\mathrm{O}(1,3)$ *is the product of an even number* ≤ 4 *of symmetries, any inversion is the product of an odd number* ≤ 4 *of symmetries* [11].

A rotation is a proper transformation of a determinant equal to 1; an inversion is an improper transformation of a determinant equal to -1.

Proper orthochronous Lorentz transformation

Consider an even number of time symmetries $(s, r, \ldots)$ and of space symmetries $(t, u, \ldots)$; one obtains

$$\begin{aligned} x' &= (utsr)x(rstu) \\ &= axa_c \end{aligned}$$

with $a = utsr \in C^+$, $a_c = r_c s_c t_c u_c = (-r)(-s)(-t)(-u) = rstu$ and $aa_c = 1$.

Other transformations

Let n be the number of time symmetries and p the number of space symmetries; their combinations give the following Lorentz transformations :

1. n odd, p odd: proper antichronous rotation
$$x' = -axa_c, \qquad aa_c = -1, \qquad a \in C^+;$$
2. n even, p odd: improper orthochronous inversion
$$x' = axa_c, \qquad aa_c = -1, \qquad a \in C^-;$$
3. n odd, p even: improper antichronous inversion
$$x' = -axa_c, \qquad aa_c = 1, \qquad a \in C^-.$$

Group O(1, 3): recapitulative table

The entire set of Lorentz transformations L is given in the table below where n is the number of time symmetries and p the number of space symmetries.

	Rotation ($\det L = 1$)	Inversion ($\det L = -1$)
orthochronous	n even, p even $x' = axa_c$ $(aa_c = 1,\ a \in C^+)$	n even, p odd $x' = axa_c = -ax_ca_c$ $(aa_c = -1,\ a \in C^-)$
antichronous	n odd, p odd $x' = -axa_c$ $(aa_c = -1,\ a \in C^+)$	n odd, p even $x' = -ax_ca_c = ax_ca_c$ $(aa_c = 1,\ a \in C^-)$

5.2 Proper orthochronous Lorentz group

5.2.1 Rotation group SO(3)

A subgroup of the proper orthochronous Lorentz group is the rotation group SO(3)

$$x' = rxr_c \tag{5.1}$$

with $r = \cos\frac{\theta}{2} + \mathbf{u}\sin\frac{\theta}{2}$, $\mathbf{u} = u^1I + u^2J + u^3K$, $(u^1)^2 + (u^2)^2 + (u^3)^2 = 1$, $x = jx^0 + k\mathbf{x}$, $x' = jx'^0 + k\mathbf{x}'$, $rr_c = 1$. One verifies that x' belongs indeed to the vector space of vectors

$$x'_c = rx_cr_c = -x'.$$

In matrix form, equation (5.1) can be written with

$$X' = \begin{bmatrix} x'^0 \\ x'^1 \\ x'^2 \\ x'^3 \end{bmatrix}, \qquad X = \begin{bmatrix} x^0 \\ x^1 \\ x^2 \\ x^3 \end{bmatrix}, \qquad x^i, x'^i \in \mathbb{R},$$

$$X' = AX,$$

$$A = \begin{bmatrix} 1 & 0 & 0 & 0 \\ 0 & a & f & g \\ 0 & m & b & h \\ 0 & n & p & c \end{bmatrix},$$

with

$$a = \left(u^1\right)^2 + \left[\left(u^2\right)^2 + \left(u^3\right)^2\right]\cos\theta,$$

$$b = \left(u^2\right)^2 + \left[\left(u^1\right)^2 + \left(u^3\right)^2\right]\cos\theta,$$

$$c = \left(u^3\right)^2 + \left[\left(u^1\right)^2 + \left(u^2\right)^2\right]\cos\theta,$$

$$\begin{aligned} f &= u^1u^2(1-\cos\theta) - u^3\sin\theta, & m &= u^1u^2(1-\cos\theta) + u^3\sin\theta, \\ g &= u^1u^3(1-\cos\theta) + u^2\sin\theta, & n &= u^1u^3(1-\cos\theta) - u^2\sin\theta, \\ h &= u^2u^3(1-\cos\theta) - u^1\sin\theta, & p &= u^2u^3(1-\cos\theta) + u^1\sin\theta. \end{aligned}$$

One obtains the same expression as with quaternions, despite the distinct nature of r and x ($r \in C^+$, $x \in C^-$). All considerations of Chapter 2 on the subgroups of SO(3) that is of r, apply here; it suffices to replace i, j, k by I, J, K respectively. For an infinitesimal rotation, one has with $r \simeq 1 + \mathbf{u}\frac{d\theta}{2}$,

$$\begin{aligned} x' &= r x r_c \\ &= \left(1 + \mathbf{u}\frac{d\theta}{2}\right) x \left(1 - \mathbf{u}\frac{d\theta}{2}\right) \\ &= x + \frac{d\theta}{2}(\mathbf{u}x - x\mathbf{u}) \\ &= x + \frac{d\theta}{2}k(\mathbf{ux} - \mathbf{xu}) \\ &= x + d\theta k\mathbf{u} \times \mathbf{x}, \\ dx &= x' - x = d\theta k\mathbf{u} \times \mathbf{x}. \end{aligned}$$

In matrix notation

$$dX = d\theta u^i M_i X, \qquad i \in (1,2,3)$$

with

$$M_1 = \begin{bmatrix} 0 & 0 & 0 & 0 \\ 0 & 0 & 0 & 0 \\ 0 & 0 & 0 & -1 \\ 0 & 0 & 1 & 0 \end{bmatrix}, \quad M_2 = \begin{bmatrix} 0 & 0 & 0 & 0 \\ 0 & 0 & 0 & 1 \\ 0 & 0 & 0 & 0 \\ 0 & -1 & 0 & 0 \end{bmatrix}, \quad M_3 = \begin{bmatrix} 0 & 0 & 0 & 0 \\ 0 & 0 & -1 & 0 \\ 0 & 1 & 0 & 0 \\ 0 & 0 & 0 & 0 \end{bmatrix}$$

and the relations

$$[M_i, M_j] = \epsilon_{ijk} M_k.$$

5.2.2 Pure Lorentz transformation

A pure Lorentz transformation is given by

$$\begin{aligned} x' &= b x b_c, \\ b &= \cosh\frac{\varphi}{2} + i\mathbf{v}\sinh\frac{\varphi}{2} \end{aligned}$$

with $\mathbf{v} = v^1 I + v^2 J + v^3 K$, $(v^1)^2 + (v^2)^2 + (v^3)^2 = 1$, $x = jx^0 + k\mathbf{x}$, $x' = jx'^0 + k\mathbf{x}'$, $bb_c = 1$. One verifies that one has indeed $x'_c = bx_c b_c = -x'$. In matrix formulation, one has with

$$X' = \begin{bmatrix} x'^0 \\ x'^1 \\ x'^2 \\ x'^3 \end{bmatrix}, \qquad X = \begin{bmatrix} x^0 \\ x^1 \\ x^2 \\ x^3 \end{bmatrix}, \qquad x^i, x'^i \in \mathbb{R},$$

$$X' = BX,$$

$$B = \begin{bmatrix} \cosh\varphi & v^1\sinh\varphi & v^2\sinh\varphi & v^3\sinh\varphi \\ v^1\sinh\varphi & a & f & g \\ v^2\sinh\varphi & f & b & h \\ v^3\sinh\varphi & g & h & c \end{bmatrix},$$

with

$$\begin{aligned} a &= 1 + \left(v^1\right)^2(\cosh\varphi - 1), & f &= v^1v^2(\cosh\varphi - 1), \\ b &= 1 + \left(v^2\right)^2(\cosh\varphi - 1), & g &= v^1v^3(\cosh\varphi - 1), \\ c &= 1 + \left(v^3\right)^2(\cosh\varphi - 1), & h &= v^2v^3(\cosh\varphi - 1). \end{aligned}$$

One will notice that the matrix is real and symmetric. For a pure infinitesimal Lorentz transformation, one obtains with $b \simeq 1 + i\mathbf{v}\frac{d\varphi}{2}$ and i anticommuting with x,

$$\begin{aligned} x' &= bxb_c \\ &= \left(1 + i\mathbf{v}\frac{d\varphi}{2}\right)x\left(1 - i\mathbf{v}\frac{d\varphi}{2}\right) \\ &= x + i\frac{d\varphi}{2}(\mathbf{v}x + x\mathbf{v}) \\ &= x + i\frac{d\varphi}{2}\left[\mathbf{v}\left(jx^0 + k\mathbf{x}\right) + \left(jx^0 + k\mathbf{x}\right)\mathbf{v}\right] \\ &= x + d\varphi\left[k\mathbf{v}x^0 - j\left(v^1x^1 + v^2x^2 + v^3x^3\right)\right], \\ dx &= x' - x = d\varphi\left[k\mathbf{v}x^0 - j\left(v^1x^1 + v^2x^2 + v^3x^3\right)\right]. \end{aligned}$$

In matrix form, one has

$$dX = d\varphi v^i K_i X, \qquad i \in (1,2,3)$$

with

$$K_1 = \begin{bmatrix} 0 & 1 & 0 & 0 \\ 1 & 0 & 0 & 0 \\ 0 & 0 & 0 & 0 \\ 0 & 0 & 0 & 0 \end{bmatrix}, \quad K_2 = \begin{bmatrix} 0 & 0 & 1 & 0 \\ 0 & 0 & 0 & 0 \\ 1 & 0 & 0 & 0 \\ 0 & 0 & 0 & 0 \end{bmatrix}, \quad K_3 = \begin{bmatrix} 0 & 0 & 0 & 1 \\ 0 & 0 & 0 & 0 \\ 0 & 0 & 0 & 0 \\ 1 & 0 & 0 & 0 \end{bmatrix}$$

and the relations

$$\begin{aligned} [K_i, K_j] &= -\epsilon_{ijk}M_k, \\ [K_i, M_j] &= \epsilon_{ijk}K_k. \end{aligned}$$

5.2.3 General Lorentz transformation

Transformation of multivectors and Clifford numbers

A general Lorentz transformation for vectors is expressed by

$$x' = axa_c$$

with $a \in C^+$, $aa_c = 1$. A bivector of the type $B = x \wedge y = -\frac{(xy-yx)}{2}$ transforms as

$$\begin{aligned} B' &= -\frac{(x'y' - y'x')}{2} \\ &= a\,(x \wedge y)\,a_c \\ &= aBa_c. \end{aligned}$$

Generally, for a product of two vectors xy one has the transformation

$$\begin{aligned} x'y' &= axa_c aya_c \\ &= axya_c; \end{aligned}$$

consequently, any multivector A (and any Clifford number) which is a linear combination of such products transforms under a proper orthochronous Lorentz transformation according to the same formula

$$A' = aAa_c \qquad (a \in C^+,\ aa_c = 1).$$

Decomposition into a rotation and a pure Lorentz transformation

The decomposition $a = br$ of a general Lorentz transformation into a rotation and a pure Lorentz transformation is obtained as follows ([15],[16]). Consider the element

$$a = a_0 + \mathbf{a} + i\,(b_0 + \mathbf{b}) \in C^+, \qquad aa_c = 1;$$

let us introduce a conjugation (transforming i into $-i$),

$$\begin{aligned} \bar{a} &= a_0 + \mathbf{a} - i\,(b_0 + \mathbf{b}) \\ &= \overline{b}\overline{r} \\ &= b_c r, \\ \bar{a}_c &= r_c b. \end{aligned}$$

Let us write $d = a\bar{a}_c = brr_cb = b^2$ $(dd_c = 1)$, hence

$$\begin{aligned} 2b^2 &= 2d, \\ b^2 d &= d^2, \\ b^2 d_c &= 1 \qquad (d_c = \bar{a}a_c); \end{aligned}$$

adding, one obtains

$$\begin{aligned} b^2(2+d+d_c) &= (1+d)^2, \\ b &= \pm\frac{(1+d)}{\sqrt{2+d+d_c}} \qquad \text{(with } d = a\overline{a}_c\text{)}, \\ r &= b_c a \\ &= \pm\frac{(1+\overline{a}a_c)\,a}{\sqrt{2+d+d_c}} \\ &= \pm\frac{(a+\overline{a})}{\sqrt{2+d+d_c}}, \end{aligned}$$

with $bb_c = rr_c = 1$. For the decomposition $a = r'b'$, one obtains similarly

$$\begin{aligned} b' &= \pm\frac{(1+d')}{\sqrt{2+d'+d'_c}} \qquad \text{(with } d' = \overline{a}_c a,\ d'_c = a_c\overline{a}\text{)}, \\ r' &= ab'_c \\ &= \pm\frac{(a+\overline{a})}{\sqrt{2+d'+d'_c}}. \end{aligned}$$

Example. Let $a = \sqrt{10}K - 3iJ$ be an even Clifford number ($a \in C^+$) and such that $aa_c = 1$. Let us decompose a into a product ($a = br = r'b'$). One has

$$\overline{a} = \sqrt{10}\,K + 3iJ, \qquad d = a\overline{a}_c = 19 + 6\sqrt{10}\,iI$$

$$\text{Answer: } b = \pm\left(\sqrt{10} + 3iI\right), \qquad r = \pm K;$$

similarly

$$d' = \overline{a}_c a = 19 - 6\sqrt{10}\,iI$$

$$\text{Answer: } b' = \pm\left(\sqrt{10} - 3iI\right), \qquad r' = \pm K.$$

5.3 Group of conformal transformations

5.3.1 Definitions

The treatment of conformal transformations within $\mathbb{H}(\mathbb{C})$ extends easily to $\mathbb{H}\otimes\mathbb{H}$. The group of conformal transformations is constituted by the transformations such that if $dx \cdot dx = 0$, then $dx' \cdot dx' = 0$ with $dx = jx^0 + k\mathbf{x}$. The group contains the spacetime translations ($x' = x + d$), Lorentz transformations ($x' = axa_c$), dilatations ($x' = \lambda x$) and the transformations

$$(x')^{-1} = x^{-1} + a_c \tag{5.2}$$

where a is a constant (four-)vector with $x^{-1} = \frac{x_c}{xx_c}$. The above equation can also be written

$$x' = \left[(x')^{-1}\right]^{-1} = \left(x^{-1} + a_c\right)^{-1}$$
$$= \left[x^{-1}\left(1 + xa_c\right)\right]^{-1} = \left[\left(1 + a_c x\right)x^{-1}\right]^{-1}$$

and using relation $(AB)^{-1} = B^{-1}A^{-1}$ which is true for any invertible Clifford number

$$x' = \left(1 + xa_c\right)^{-1} x \tag{5.3}$$
$$= x\left(1 + a_c x\right)^{-1}. \tag{5.4}$$

Equivalently, one has

$$x' = \left(\frac{x_c}{xx_c} + a_c\right)^{-1} = \frac{\left(\frac{x}{xx_c} + a\right)}{\left(\frac{x_c}{xx_c} + a_c\right)\left(\frac{x}{xx_c} + a\right)},$$
$$x' = \frac{x + a\,(x \cdot x)}{1 + 2(x \cdot a) + (a \cdot a)(x \cdot x)}.$$

The inverse transformation is given by

$$x^{-1} = (x')^{-1} - a_c$$

hence

$$x = \left(1 - x'a_c\right)^{-1} x' \tag{5.5}$$
$$= x'\left(1 - a_c x'\right)^{-1}. \tag{5.6}$$

5.3.2 Properties of conformal transformations

For any invertible Clifford number A ($AA^{-1} = A^{-1}A = 1$), one has

$$dA^{-1}A + A^{-1}dA = 0,$$
$$dA^{-1} = -A^{-1}dAA^{-1}.$$

Differentiating equation (5.2), one obtains

$$d\left[(x')^{-1}\right] = d\left(x^{-1}\right), \tag{5.7}$$
$$(x')^{-1} dx' (x')^{-1} = x^{-1}dxx^{-1}, \tag{5.8}$$
$$\frac{x'_c dx' x'_c}{(x'x'_c)^2} = \frac{x_c dx x_c}{(xx_c)^2}; \tag{5.9}$$

multiplying by the conjugate equation, one has

$$\frac{x'_c dx' x'_c x' dx'_c x'}{(x' x'_c)^4} = \frac{x_c dx x_c x dx_c x}{(x x_c)^4}, \tag{5.10}$$

$$dx' dx'_c = \frac{(x'.x')^2}{(x.x)^2} dx dx_c \tag{5.11}$$

which shows that the transformation is indeed a conformal transformation. Equation (5.8) gives

$$(x')^{-1} dx' (x')^{-1} = x^{-1} dx x^{-1},$$
$$dx' = x' x^{-1} dx x^{-1} x',$$

and using equations (5.3), (5.4), one obtains the relation

$$dx' = (1 + x a_c)^{-1} dx (1 + a_c x)^{-1}. \tag{5.12}$$

Finally, from equations (5.3), (5.4) one has

$$x' = (1 + x a_c)^{-1} x = x (1 + a_c x)^{-1},$$
$$(1 + x a_c) x' = x = x' (1 + a_c x),$$

hence

$$x a_c x' = x' a_c x, \tag{5.13}$$
$$x x_c = (1 + x a_c) x' x'_c (1 + a x_c), \tag{5.14}$$
$$x' x'_c = \frac{x x_c}{1 + 2(x \cdot a) + (a \cdot a)(x \cdot x)}. \tag{5.15}$$

5.3.3 Transformation of multivectors

A conformal transformation is a relation of the type $x'^i = f^i(x^0, x^1, x^2, x^3)$ with

$$dx'^i = \frac{\partial x'^i}{\partial x^k} dx^k.$$

A contravariant (four-)vector A, by definition, transforms according to the formula

$$A'^i = \frac{\partial x'^i}{\partial x^k} A^k.$$

The equation (5.12)

$$dx' = (1 + x a_c)^{-1} dx (1 + a_c x)^{-1}$$

then gives the transformation of a vector $A = jx^0 + k\mathbf{A}$,

$$A' = (1 + xa_c)^{-1} A (1 + a_c x)^{-1} . \tag{5.16}$$

For a product of two vectors A, B one obtains

$$\begin{aligned} A'B' &= (1 + xa_c)^{-1} A (1 + a_c x)^{-1} (1 + xa_c)^{-1} B (1 + a_c x)^{-1} \\ &= K (1 + xa_c)^{-1} AB (1 + a_c x)^{-1} \end{aligned}$$

with

$$K = \frac{1}{1 + 2(x \cdot a) + (a \cdot a)(x \cdot x)};$$

for a bivector $A \wedge B = -\frac{(AB-BA)}{2}$ one then has

$$A' \wedge B' = K (1 + xa_c)^{-1} (A \wedge B) (1 + a_c x)^{-1} ;$$

one verifies that $A' \wedge B'$ is indeed a bivector. Similarly, one obtains for a trivector and a pseudoscalar

$$\begin{aligned} A' \wedge B' \wedge C' &= K^2 (1 + xa_c)^{-1} (A \wedge B \wedge C) (1 + a_c x)^{-1} , \\ A' \wedge B' \wedge C' \wedge D' &= K^4 (A \wedge B \wedge C \wedge D) . \end{aligned}$$

Furthermore,

$$\begin{aligned} A'A'_c &= K^2 A A_c, \\ A' \cdot B' &= K^2 (A \cdot B) \end{aligned}$$

which shows that the conformal transformation conserves the angles.

5.4 Dirac algebra

5.4.1 Dirac equation

The Dirac algebra is isomorphic to the Clifford algebra $\mathbb{H} \otimes \mathbb{H}$ over $\mathbb{C}$ and can be represented by 2×2 matrices over complex quaternions with the following generators:

$$e_0 (\equiv j) = \begin{bmatrix} i' & 0 \\ 0 & -i' \end{bmatrix}, \qquad e_1 (\equiv kI) = \begin{bmatrix} 0 & i'i \\ i'i & 0 \end{bmatrix},$$

$$e_2 (\equiv kJ) = \begin{bmatrix} 0 & i'j \\ i'j & 0 \end{bmatrix}, \qquad e_3 (\equiv kK) = \begin{bmatrix} 0 & i'k \\ i'k & 0 \end{bmatrix},$$

where i' is the ordinary complex imaginary ($i'^2 = -1$) and $e_0^2 = -1$, $e_1^2 = e_2^2 = e_3^2 = 1$. The other matrices are given by

$$1 = \begin{bmatrix} 1 & 0 \\ 0 & 1 \end{bmatrix}, \qquad I = \begin{bmatrix} i & 0 \\ 0 & i \end{bmatrix}, \qquad J = \begin{bmatrix} j & 0 \\ 0 & j \end{bmatrix},$$

$$K = \begin{bmatrix} k & 0 \\ 0 & k \end{bmatrix}, \qquad i = \begin{bmatrix} 0 & -1 \\ 1 & 0 \end{bmatrix}, \qquad iI = \begin{bmatrix} 0 & -i \\ i & 0 \end{bmatrix},$$

$$iJ = \begin{bmatrix} 0 & -j \\ j & 0 \end{bmatrix}, \qquad iK = \begin{bmatrix} 0 & -k \\ k & 0 \end{bmatrix}, \qquad jI = \begin{bmatrix} i'i & 0 \\ 0 & -i'i \end{bmatrix},$$

$$jJ = \begin{bmatrix} i'j & 0 \\ 0 & -i'j \end{bmatrix}, \qquad jK = \begin{bmatrix} i'k & 0 \\ 0 & -i'k \end{bmatrix}, \qquad k = \begin{bmatrix} 0 & i' \\ i' & 0 \end{bmatrix}.$$

The Dirac spinor can be expressed as a left ideal

$$\Psi = AE = \begin{bmatrix} q_1 f \\ q_2 f \end{bmatrix}, \qquad E = \begin{bmatrix} f & 0 \\ 0 & 0 \end{bmatrix}$$

where E is a primitive idempotent ($E^2 = E$), $f = (1 + i'k)/2$; q_1, q_2 being real quaternions and A a general element of the algebra. The hermitian norm is expressed by

$$\Psi\Psi^\dagger = (q_1 q_{1c} + q_2 q_{2c})\, f$$

where $\Psi^\dagger$ is the transposed, quaternionic conjugated and complex conjugated matrix of Ψ. The Dirac equation in presence of an electromagnetic field is

$$(i\hbar\nabla - eA - i'mc)\,\Psi = 0$$

with the four-nabla operator $\nabla = j\frac{\partial}{c\partial t} - k\boldsymbol{\nabla}$, the vector potential $A = j\frac{V}{c} + k\mathbf{A}$ and e the electric charge of the particle.

5.4.2 Unitary and symplectic unitary groups

Having defined the Clifford algebra $\mathbb{H} \otimes \mathbb{H}$ over $\mathbb{C}$, let A be a 2×2 matrix having as elements complex quaternions $\underline{q_i}$ and its adjoint $A^\dagger$,

$$A = \begin{bmatrix} \underline{q_1} & \underline{q_2} \\ \underline{q_3} & \underline{q_4} \end{bmatrix}, \qquad A^\dagger = \begin{bmatrix} \underline{q_{1c}^*} & \underline{q_{3c}^*} \\ \underline{q_{2c}^*} & \underline{q_{4c}^*} \end{bmatrix},$$

where the sympbols $*$ and c represent respectively the complex conjugation and the quaternionic conjugation. The adjunction transforms i' and i, j, k, I, J, K into their opposites, as one can verify directly on the basis matrices. A selfadjoint Clifford number ($H = H^\dagger$) is consequently of the type

$$H = (a + i'ib + i'jc + i'kd;\ i'\mathbf{p} + i\mathbf{q} + j\mathbf{r} + k\mathbf{s})$$

with $\mathbf{p} = p^1 I + p^2 J + p^3 K$ (etc.) and real coefficients. The unitary group $\mathrm{SU}\left[2, \mathbb{H}(\mathbb{C})\right]$, isomorphic to SU(4), is the set of matrices A such that

$$AA^\dagger = A^\dagger A = 1.$$

If one restricts the matrices to real quaternion matrices, one obtains as a subgroup the symplectic unitary group $\mathrm{USp}(2, \mathbb{H})$ [39, p. 232]. The elements of the unitary group being of the type $e^{i'H}$ (with $H = H^\dagger$), one can choose for the 15 generators of SU(4),

$$\left[\begin{array}{c} \left(e^{i\theta}, e^{I\theta}, e^{J\theta}, e^{K\theta}, e^{i'jI\theta}, e^{i'jJ\theta}, e^{i'jK\theta}, e^{i'kI\theta}, e^{i'kJ\theta}, e^{i'kK\theta} \right) \\ e^{j\theta}, e^{k\theta}, e^{i'iI\theta}, e^{i'iJ\theta}, e^{i'iK\theta}. \end{array} \right], \tag{5.17}$$

where θ is a real generic parameter with the usual series development

$$\begin{aligned} e^{I\theta} &= \cos\theta + I\sin\theta, \\ e^{i'iJ\theta} &= \cos\theta + i'iJ\sin\theta \quad \text{(etc.)}. \end{aligned}$$

The 10 matrices within parentheses constitute the symplectic unitary group $\mathrm{USp}(2; \mathbb{H})$. Explicitly, one has

$$e^{i\theta} = \begin{bmatrix} \cos\theta & -\sin\theta \\ \sin\theta & \cos\theta \end{bmatrix}, \qquad e^{I\theta} = \begin{bmatrix} \cos\theta + i\sin\theta & 0 \\ 0 & \cos\theta + i\sin\theta \end{bmatrix},$$

$$e^{J\theta} = \begin{bmatrix} \cos\theta + j\sin\theta & 0 \\ 0 & \cos\theta + j\sin\theta \end{bmatrix},$$

$$e^{K\theta} = \begin{bmatrix} \cos\theta + k\sin\theta & 0 \\ 0 & \cos\theta + k\sin\theta \end{bmatrix},$$

$$e^{i'jI\theta} = \begin{bmatrix} \cos\theta - i\sin\theta & 0 \\ 0 & \cos\theta + i\sin\theta \end{bmatrix},$$

$$e^{i'jJ\theta} = \begin{bmatrix} \cos\theta - j\sin\theta & 0 \\ 0 & \cos\theta + j\sin\theta \end{bmatrix},$$

$$e^{i'jK\theta} = \begin{bmatrix} \cos\theta - k\sin\theta & 0 \\ 0 & \cos\theta + k\sin\theta \end{bmatrix},$$

$$e^{i'kI\theta} = \begin{bmatrix} \cos\theta & -i\sin\theta \\ -i\sin\theta & \cos\theta \end{bmatrix}, \qquad e^{i'kJ\theta} = \begin{bmatrix} \cos\theta & -j\sin\theta \\ -j\sin\theta & \cos\theta \end{bmatrix},$$

$$e^{i'kK\theta} = \begin{bmatrix} \cos\theta & -k\sin\theta \\ -k\sin\theta & \cos\theta \end{bmatrix}.$$

One verifies that these matrices have indeed real quaternions as coefficients. The other elements of the SU(4) group are given by

$$e^{j\theta} = \begin{bmatrix} \cos\theta + i'\sin\theta & 0 \\ 0 & \cos\theta - i'\sin\theta \end{bmatrix}, \qquad e^{k\theta} = \begin{bmatrix} \cos\theta & i'\sin\theta \\ i'\sin\theta & \cos\theta \end{bmatrix},$$

$$e^{i'iI\theta} = \begin{bmatrix} \cos\theta & -i'\sin\theta \\ i'\sin\theta & \cos\theta \end{bmatrix}, \qquad e^{i'iJ\theta} = \begin{bmatrix} \cos\theta & -i'j\sin\theta \\ i'j\sin\theta & \cos\theta \end{bmatrix},$$

$$e^{i'ik\theta} = \begin{bmatrix} \cos\theta & -i'\sin\theta \\ -i'\sin\theta & \cos\theta \end{bmatrix}.$$

To obtain a representation in terms of 4×4 complex matrices, one can choose for the generators of the Clifford algebra over $\mathbb{C}$

$$e_0(\equiv j) = \begin{bmatrix} i' & 0 & 0 & 0 \\ 0 & i' & 0 & 0 \\ 0 & 0 & -i' & 0 \\ 0 & 0 & 0 & -i' \end{bmatrix}, \qquad e_1(\equiv kI) = \begin{bmatrix} 0 & 0 & 0 & i' \\ 0 & 0 & i' & 0 \\ 0 & -i' & 0 & 0 \\ -i' & 0 & 0 & 0 \end{bmatrix},$$

$$e_2(\equiv kJ) = \begin{bmatrix} 0 & 0 & 0 & 1 \\ 0 & 0 & -1 & 0 \\ 0 & -1 & 0 & 0 \\ 1 & 0 & 0 & 0 \end{bmatrix}, \qquad e_3(\equiv kK) = \begin{bmatrix} 0 & 0 & i' & 0 \\ 0 & 0 & 0 & -i' \\ -i' & 0 & 0 & 0 \\ 0 & i' & 0 & 0 \end{bmatrix},$$

with $e_0^2 = -1$, $e_1^2 = e_2^2 = e_3^2 = 1$.

5.5 Exercises

E5-1 Consider a special pure Lorentz transformation (b) along the Ox axis (velocity $v = \frac{c}{3}$) followed by a rotation (r) of $\frac{\pi}{4}$ around the same axis. Express the resulting Lorentz transformation $X = aX'a_c$.

E5-2 Consider a special Lorentz transformation (b_1) along the Ox axis (velocity $v = \frac{c}{3}$) followed by a pure Lorentz transformation (b_2) along the Oy axis (with $v = \frac{c}{3}$). Express the resulting Lorentz transformation $X = aX'a_c$ with $a = b_2 b_1$. Decompose the resulting Lorentz transformation into a rotation followed by a pure Lorentz transformation ($a = br$). Give the direction of the Lorentz transformation, the velocity and the angle of rotation.

E5-3 Consider an orthonormal system of axes (O, x, y, z) and a cube with vertices $A(0,0,0)$, $B(1,0,0)$, $C(1,1,0)$, $D(0,1,0)$, $E(0,1,1)$, $F(0,0,1)$, $G(1,0,1)$, $H(1,1,1)$. Determine the transform of the vertices of this cube under a conformal transformation

$$x' = (1 + xa_c)^{-1}x$$

with $a = 2kI + kK$, $x = ctj + k\mathbf{x}$, $x' = ct'j + k\mathbf{x}'$.

E5-4 Consider the matrices

$$A = \begin{bmatrix} 1 & j \\ k & i \end{bmatrix}, \qquad B = \begin{bmatrix} 1 & k \\ i & j \end{bmatrix}, \qquad C = \begin{bmatrix} 1 & i \\ k & j \end{bmatrix}.$$

Determine AB, BA, A^{-1}, B^{-1}, $(AB)^{-1}$, $(BA)^{-1}$; does the matrix C^{-1} exist?

Chapter 6

Special relativity

This chapter develops the special theory of relativity within the Clifford algebra $\mathbb{H}\otimes\mathbb{H}$ over $\mathbb{R}$. The relativistic kinematics and relativistic dynamics of a point mass are examined.

6.1 Lorentz transformation

6.1.1 Special Lorentz transformation

Consider a reference frame $K(O,x,y,z)$ at rest and a reference frame $K'(O',x',y',z')$ moving along the Ox axis with a constant velocity $\mathbf{v}$ (Figure 6.1). With $X = jx^0 + k\mathbf{x}$, $X' = jx'^0 + k\mathbf{x}'$ ($x^0 = ct$, $\mathbf{x} = xI + yJ + zK$), $b = \cosh\frac{\varphi}{2} - iI\sinh\frac{\varphi}{2}$, the Lorentz transformation is expressed by

$$X' = bXb_c.$$

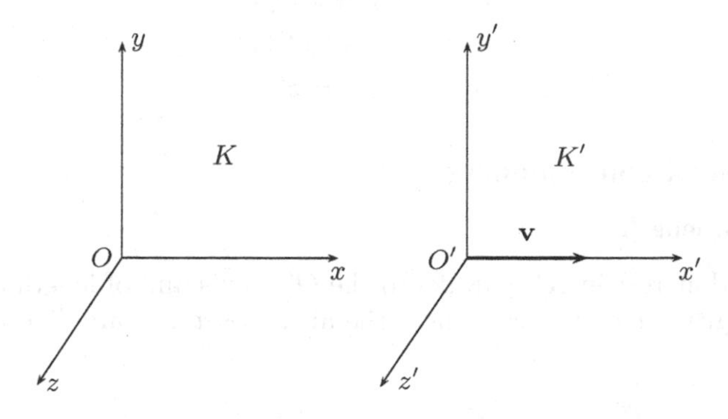

Figure 6.1: Special Lorentz transformation (pure).

Explicitly, one obtains

$$
\begin{aligned}
x'^0 &= x^0 \cosh\varphi - x' \sinh\varphi, \\
x' &= -x^0 \sinh\varphi + x' \cosh\varphi, \\
y' &= y, \quad z' = z.
\end{aligned}
$$

Writing $\tanh\varphi = \frac{v}{c} = \beta$, $\gamma = \cosh\varphi$ with

$$
\begin{aligned}
\gamma^2 &= 1 + \sinh^2\varphi = 1 + \gamma^2\beta^2, \\
\gamma &= \frac{1}{\sqrt{1-\beta^2}},
\end{aligned}
$$

the transformation becomes

$$
\begin{aligned}
ct' &= \gamma\left(ct - x\beta\right), \\
x' &= \gamma\left(x - \beta ct\right), \\
y' &= y, \quad z' = z.
\end{aligned}
$$

Taking $\beta = \frac{v}{c} \ll 1$ $(\gamma \simeq 1)$, one obtains the Galilean transformation

$$
\begin{aligned}
t' &= t, \\
x' &= x - vt, \\
y' &= y, \quad z' = z.
\end{aligned}
$$

The inverse transformation is expressed by

$$
X = b_c X' b
$$

with $b_c = \cosh\frac{\varphi}{2} + iI\sinh\frac{\varphi}{2}$, hence

$$
\begin{aligned}
ct &= \gamma\left(ct' + x'\beta\right), \\
x &= \gamma\left(x' + \beta ct'\right), \\
y &= y', \quad z = z'.
\end{aligned}
$$

6.1.2 Physical consequences

Contraction of length

Consider a rod at rest in K', parallel to the $O'x'$ axis and of length $l'_0 = x'_2 - x'_1$. In K, the length is $l = x_2 - x_1$, where the abscisses x_2, x_1 are determined at the same time t:

$$
\begin{aligned}
x'_2 &= \gamma\left(x_2 - \beta ct\right), \\
x'_1 &= \gamma\left(x_1 - \beta ct\right),
\end{aligned}
$$

hence

$$l = x_2 - x_1 = \frac{x_2' - x_1'}{\gamma}$$
$$= l_0' \sqrt{1 - \beta^2} \le l_0';$$

the observer in K concludes to a contraction. Reciprocally, let $l_0 = x_2 - x_1$ be the length of a rod parallel to the Ox axis in K. The observer in K' measures at the same time t',

$$x_2 = \gamma \left(x_2' + \beta c t' \right),$$
$$x_1 = \gamma \left(x_1' + \beta c t' \right),$$

hence

$$l' = x_2' - x_1' = \frac{x_2 - x_1}{\gamma}$$
$$= l_0 \sqrt{1 - \beta^2} \le l_0;$$

the observer in K' concludes also to a contraction.

Example. Take $v = 300$ km/s, $\beta = 10^{-3}$, $\beta^2 = 10^{-6}$, $\sqrt{1-\beta^2} = 1 - 5 \cdot 10^{-7}$, the relative variation of l is $5 \cdot 10^{-7}$.

Time dilatation

A time interval $\Delta t' = t_2' - t_1'$ measured at the same point $\Delta x' = x_2' - x_1' = 0$ of K' corresponds in K to a time interval $\Delta t = t_2 - t_1$ with

$$c t_2 = \gamma \left(c t_2' + x_2' \beta \right),$$
$$c t_1 = \gamma \left(c t_1' + x_1' \beta \right),$$

hence

$$\Delta t = t_2 - t_1 = \gamma \, \Delta \, t'$$
$$= \frac{\Delta t'}{\sqrt{1 - \beta^2}} \ge \Delta t'.$$

The observer in K concludes to a slowing down of physical phenomena. Reciprocally, a time interval $\Delta t = t_2 - t_1$ measured at the same point $x_2 = x_1$ in K, gives in K',

$$c t_2' = \gamma \left(c t_2 - x_2 \beta \right),$$
$$c t_1' = \gamma \left(c t_1 - x_1 \beta \right),$$
$$\Delta t' = t_2' - t_1'$$
$$= \frac{\Delta t}{\sqrt{1 - \beta^2}} \ge \Delta t.$$

Hence, the conclusion is the same.

6.1.3 General Lorentz transformation

For a general Lorentz transformation, one has $a = br$ (or $a' = rb$) with $r = \cos\frac{\theta}{2} + \mathbf{u}\sin\frac{\theta}{2}$, $b = \cosh\frac{\varphi}{2} + i\mathbf{v}\sinh\frac{\varphi}{2}$ ($\mathbf{u}\cdot\mathbf{u} = \mathbf{v}\cdot\mathbf{v} = 1$). Explicitly, one obtains

$$a = br = \begin{bmatrix} \cos\frac{\theta}{2}\cosh\frac{\varphi}{2} + \mathbf{u}\cosh\frac{\varphi}{2}\sin\frac{\theta}{2} \\ -i\mathbf{u}\cdot\mathbf{v}\sin\frac{\theta}{2}\sinh\frac{\varphi}{2} \\ +i\left(\mathbf{v}\cos\frac{\theta}{2} - \mathbf{u}\times\mathbf{v}\sin\frac{\theta}{2}\sinh\frac{\varphi}{2}\right) \end{bmatrix},$$

$$a' = rb = \begin{bmatrix} \cos\frac{\theta}{2}\cosh\frac{\varphi}{2} + \mathbf{u}\cosh\frac{\varphi}{2}\sin\frac{\theta}{2} \\ -i\mathbf{u}\cdot\mathbf{v}\sin\frac{\theta}{2}\sinh\frac{\varphi}{2} \\ +i\left(\mathbf{v}\cos\frac{\theta}{2} + \mathbf{u}\times\mathbf{v}\sin\frac{\theta}{2}\sinh\frac{\varphi}{2}\right) \end{bmatrix}$$

with $aa_c = a'a'_c = 1$. The general Lorentz transformation is simply expressed by

$$X' = aXa_c \qquad (\text{or } a'Xa'_c)$$

with $X = jx^0 + k\mathbf{x}$, $X' = jx'^0 + k\mathbf{x}'$.

6.2 Relativistic kinematics

6.2.1 Four-vectors

Transformation of a four-vector

An arbitrary four-vector $A = ja^0 + k\mathbf{a}$ transforms under a special Lorentz transformation as

$$A' = bAb_c$$

with $b = \cosh\frac{\varphi}{2} - iI\sinh\frac{\varphi}{2}$; explicitly, writing

$$\gamma = \cosh\varphi = \frac{1}{\sqrt{1-\frac{v^2}{c^2}}}, \qquad \tanh\varphi = \frac{v}{c},$$

one has

$$a'^0 = \gamma\left(a^0 - a^1\frac{v}{c}\right),$$
$$a'^1 = \gamma\left(a^1 - a^0\frac{v}{c}\right),$$
$$a'^2 = a^2, \quad a'^3 = a^3;$$

reciprocally, one has

$$A = b_c A' b$$

which yields

$$a^0 = \gamma\left(a'^0 + a'^1\frac{v}{c}\right),$$
$$a^1 = \gamma\left(a'^1 + a'^0\frac{v}{c}\right),$$
$$a^2 = a'^2, \quad a^3 = a'^3.$$

If one considers a general Lorentz transformation, one obtains the following formulas ([40, p. 134], [53, p.123]) with $b = \cosh\frac{\varphi}{2} - i\frac{\mathbf{v}}{v}\sinh\frac{\varphi}{2}$ where $\mathbf{v}$ is the velocity (of norm v) of the reference frame K' with respect to the reference frame K

$$A' = bAb_c$$

or explicitly

$$a'^0 = \gamma\left[a^0 - \left(\mathbf{a}\cdot\frac{\mathbf{v}}{v}\right)\frac{v}{c}\right],$$
$$\mathbf{a}' = \mathbf{a} + \frac{\mathbf{v}}{v}\left[\left(\mathbf{a}\cdot\frac{\mathbf{v}}{v}\right)(\gamma-1) - a^0\gamma\frac{v}{c}\right];$$

the reciprocal formulas are

$$A = b_c A' b,$$

$$a^0 = \gamma\left[a'^0 + \left(\mathbf{a}'\cdot\frac{\mathbf{v}}{v}\right)\frac{v}{c}\right],$$
$$\mathbf{a} = \mathbf{a}' + \frac{\mathbf{v}}{v}\left[\left(\mathbf{a}'\cdot\frac{\mathbf{v}}{v}\right)(\gamma-1) + a'^0\gamma\frac{v}{c}\right].$$

Four-velocity

Let $X = jct + k\mathbf{x}$ be the spacetime four-vector of a particle with $dX = jcdt + k\mathbf{dx}$ and the relativistic invariant

$$dXdX_c = c^2dt^2 - (\mathbf{dx})^2 = c^2dt^2\left[1 - \frac{v^2}{c^2}\right]$$
$$= \frac{c^2dt^2}{\gamma^2} = c^2d\tau^2$$

which defines the proper time

$$d\tau = \frac{dt}{\gamma} = dt\sqrt{1 - \frac{v^2}{c^2}}$$

with $\gamma = \frac{1}{\sqrt{1-\frac{v^2}{c^2}}}$, v being the velocity of the particle. The four-velocity V is defined by

$$V = \frac{dX}{d\tau} = j\gamma c + k\gamma\mathbf{v}$$

with $\mathbf{v} = \frac{d\mathbf{x}}{dt}$ and $VV_c = c^2$; V can also be written in the form

$$V = c\,(j\cosh\theta + k\mathbf{m}\sinh\theta)$$

with $\mathbf{v} = v\mathbf{m}$ $(\mathbf{m}\cdot\mathbf{m} = 1)$ and $\tanh\theta = \frac{v}{c}$, $\gamma = \cosh\theta$, $\sinh\theta = \gamma\frac{v}{c}$.

Four-acceleration

Let $V = j\gamma c + k\gamma\mathbf{v}$ be the four-velocity and $d\tau = \frac{dt}{\gamma}$ the proper time differential; the four-acceleration is defined by

$$A = \frac{dV}{d\tau} = \gamma\left[jc\dot{\gamma} + k\left(\dot{\gamma}\mathbf{v} + \gamma\mathbf{a}\right)\right]$$

with

$$\dot{\gamma} = \frac{d\gamma}{dt} = \frac{v\dot{v}}{c^2}\gamma^3 \tag{6.1}$$

$$= (\mathbf{v}\cdot\mathbf{a})\,\frac{\gamma^3}{c^2} \tag{6.2}$$

where the relation $\mathbf{v}\cdot\dot{\mathbf{v}} = v\dot{v}$ deduced from $(\mathbf{v})^2 = v^2$ has been used. Furthermore, $V\cdot V = c^2$ from which one obtains by differentiating with respect to τ, $V\cdot A = 0$ which gives again equation (6.2)

$$(\mathbf{v}\cdot\mathbf{a}) = c^2\frac{\dot{\gamma}}{\gamma^3}.$$

The four-acceleration can be written

$$A = j\gamma^4\frac{(\mathbf{v}\cdot\mathbf{a})}{c} + k\left[\gamma^4\frac{(\mathbf{v}\cdot\mathbf{a})\,\mathbf{v}}{c^2} + \gamma^2\mathbf{a}\right]$$

with $\gamma = \frac{1}{\sqrt{1-\frac{v^2}{c^2}}}$; one has the relativistic invariant

$$\begin{aligned} AA_c &= -\gamma^6\frac{(\mathbf{v}\cdot\mathbf{a})^2}{c^2} - \gamma^4\,(\mathbf{a})^2 \\ &= \gamma^6\left[-a^2 + \frac{\mathbf{v}\times\mathbf{a}}{c^2}\right] \end{aligned}$$

with $(\mathbf{v}\times\mathbf{a})^2 = v^2a^2 - (\mathbf{v}\cdot\mathbf{a})^2$. The bivector $V\wedge A$ is given by

$$V\wedge A = (\mathbf{v}\times\mathbf{a})\,\gamma^3 - i\mathbf{a}c\gamma^3;$$

since $VA = -(V\cdot A) - (V\wedge A) = -(V\wedge A)$, one has

$$\begin{aligned} (V\wedge A)\,(V\wedge A)_c &= VAA_cV_c = c^2AA_c \\ &= \gamma^6\left[-a^2c^2 + (\mathbf{v}\times\mathbf{a})^2\right]. \end{aligned}$$

The four-acceleration in the proper frame ($\mathbf{v} = 0$) is simply $A = k\mathbf{a}$ with

$$AA_c = -(\mathbf{a}_p)^2 = -a_p^2.$$

When $\mathbf{a}$ is parallel to $\mathbf{v}$, one has

$$AA_c = -\gamma^6 a^2 = -a_p^2,$$
$$a_p = \gamma^3 a.$$

If $\mathbf{a}$ is perpendicular to $\mathbf{v}$, one has

$$AA_c = -\gamma^6 a^2 \left[1 - \frac{v^2}{c^2}\right]$$
$$= -\gamma^4 a^2 = -a_p^2,$$
$$a_p = \gamma^2 a;$$

when $\gamma \gg 1$, one remarks that the proper acceleration can be very much larger than the acceleration in the laboratory frame [42, p. 101].

Wave four-vector

The wave four-vector is defined with $\mathbf{k} = \frac{2\pi}{\lambda}\mathbf{n}$ by

$$K = j\frac{\omega}{c} + k\mathbf{k};$$

writing $X = jct + k\mathbf{r}$ one has the relativistic invariants

$$KK_c = \frac{\omega^2}{c^2} - (\mathbf{k})^2,$$
$$K \cdot X = \omega t - \mathbf{k} \cdot \mathbf{r}.$$

6.2.2 Addition of velocities

Special Lorentz transformation

Consider the special Lorentz transformation

$$V' = bVb_c \tag{6.3}$$

with

$$b = \cosh\frac{\varphi}{2} - iI\sinh\frac{\varphi}{2}, \qquad \tanh\varphi = \frac{w}{c}, \qquad \cosh\varphi = \gamma = \frac{1}{\sqrt{1 - \frac{w^2}{c^2}}},$$

the four-velocities

$$V = c\left(j\cosh\theta + k\mathbf{m}\sinh\theta\right), \qquad V' = c\left(j\cosh\theta' + k\mathbf{m}'\sinh\theta'\right),$$

and

$$\tanh\theta = \frac{v}{c}, \qquad \tanh\theta' = \frac{v'}{c},$$
$$\mathbf{v} = v\mathbf{m}, \qquad \mathbf{v}' = v'\mathbf{m}' \qquad (\mathbf{m}\cdot\mathbf{m} = \mathbf{m}'\cdot\mathbf{m}' = 1).$$

Equations (6.3) read

$$\cosh\theta' = \cosh\theta\cosh\varphi - m^1\sinh\theta\sinh\varphi, \tag{6.4}$$
$$m'^1\sinh\theta' = m^1\cosh\varphi\sinh\theta - \cosh\theta\sinh\varphi, \tag{6.5}$$
$$m'^2\sinh\theta' = m^2\sinh\theta, \tag{6.6}$$
$$m'^3\sinh\theta' = m^3\sinh\theta. \tag{6.7}$$

Dividing equation (6.5) by the equation (6.4), one obtains

$$m'^1\tanh\theta' = \frac{m^1\tanh\theta - \tanh\varphi}{1 - m^1\tanh\theta\tanh\varphi}$$

and thus

$$v'^1 = \frac{v^1 - w}{1 - \dfrac{v^1 w}{c^2}}. \tag{6.8}$$

Similarly, one obtains

$$v'^2 = \frac{v^2}{\cosh\varphi\left(1 - \dfrac{v^1 w}{c^2}\right)}, \tag{6.9}$$
$$v'^3 = \frac{v^3}{\cosh\varphi\left(1 - \dfrac{v^1 w}{c^2}\right)} \tag{6.10}$$

with $\cosh\varphi = \gamma = \frac{1}{\sqrt{1-\frac{w^2}{c^2}}}$. The inverse formulas are (with $\beta = \frac{w}{c}$)

$$v^1 = \frac{v'^1 + w}{1 + \dfrac{v'^1 w}{c^2}}, \tag{6.11}$$
$$v^2 = \frac{v'^2\sqrt{1-\beta^2}}{1 + \dfrac{v'^1 w}{c^2}}, \tag{6.12}$$
$$v^3 = \frac{v'^3\sqrt{1-\beta^2}}{1 + \dfrac{v'^1 w}{c^2}}. \tag{6.13}$$

General pure Lorentz transformation

It is expressed for four-velocities by

$$V' = bVb_c \tag{6.14}$$

with $b = \left(\cosh\frac{\varphi}{2} - i\frac{\mathbf{w}}{w}\sinh\frac{\varphi}{2}\right)$, $\mathbf{v} = v\mathbf{m}$, $\mathbf{v}' = v'\mathbf{m}'$ ($\mathbf{m}\cdot\mathbf{m} = \mathbf{m}'\cdot\mathbf{m}' = 1$), $\tanh\varphi = \frac{w}{c}$, $\tanh\theta = \frac{v}{c}$, $\tanh\theta' = \frac{v'}{c}$. Explicitly, equation (6.14) reads

$$\cosh\theta' = \cosh\theta\cosh\varphi\left[1 - \left(\mathbf{m}\cdot\frac{\mathbf{w}}{w}\right)\tanh\theta\tanh\varphi\right], \tag{6.15}$$

$$\mathbf{m}'\sinh\theta' = \mathbf{m}\sinh\theta - \frac{\mathbf{w}}{w}\left(\mathbf{m}\cdot\frac{\mathbf{w}}{w}\right)\sinh\theta \tag{6.16}$$

$$+\frac{\mathbf{w}}{w}\left(\mathbf{m}\cdot\frac{\mathbf{w}}{w}\right)\sinh\theta\cosh\varphi - \frac{\mathbf{w}}{w}\cosh\theta\sinh\varphi. \tag{6.17}$$

Dividing equation (6.17) by equation (6.15), one obtains with $\beta = \frac{w}{c}$ the formulas [40, p. 75]

$$\mathbf{v}' = \frac{\mathbf{v}\sqrt{1-\beta^2} + \mathbf{w}\left[\frac{\mathbf{v}\cdot\mathbf{w}}{w^2}\left(1-\sqrt{1-\beta^2}\right) - 1\right]}{1 - \frac{\mathbf{v}\cdot\mathbf{w}}{c^2}}$$

and the inverse formula

$$\mathbf{v} = \frac{\mathbf{v}'\sqrt{1-\beta^2} + \mathbf{w}\left[\frac{\mathbf{v}'\cdot\mathbf{w}}{w^2}\left(1-\sqrt{1-\beta^2}\right) + 1\right]}{1 + \frac{\mathbf{v}'\cdot\mathbf{w}}{c^2}}.$$

6.3 Relativistic dynamics of a point mass

6.3.1 Four-momentum

Let $V = j\gamma c + k\gamma\mathbf{v}$ be the four-velocity of a particle and m_0 its mass, the four-momentum is defined by

$$\begin{aligned} P &= m_0 V = j m_0\gamma c + k\gamma m_0\mathbf{v} \\ &= j\frac{E}{c} + k\mathbf{p} \end{aligned}$$

where $E = \gamma m_0 c^2$ is the energy of the particle and $\mathbf{p} = \gamma m_0\mathbf{v}$ its momentum (of norm p). Furthermore, one has the relativistic invariant

$$\begin{aligned} PP_c &= \frac{E^2}{c^2} - (\mathbf{p})^2 \\ &= m_0^2 V V_c = m_0^2 c^2, \end{aligned}$$

hence

$$E^2 = p^2c^2 + m_0^2c^4.$$

In the proper frame ($\mathbf{v} = 0$), one has $E_0 = m_0c^2$; the kinetic energy is defined by

$$T = m_0c^2\left(\gamma - 1\right).$$

Under a standard Lorentz transformation, the four-momentum transforms as $P' = bPb_c$ with $b = \left(\cosh\frac{\varphi}{2} - iI\sinh\frac{\varphi}{2}\right)$ and $\tanh\varphi = \frac{w}{c} = \beta$, $\gamma = \frac{1}{\sqrt{1-\frac{w^2}{c^2}}}$

$$\begin{aligned} E' &= \gamma\left(E - p^1 w\right), \\ p'^1 &= \gamma\left(p^1 - \frac{E}{c^2}w\right), \\ p'^2 &= p^2, \quad p'^3 = p^3. \end{aligned}$$

The inverse transformation is $P' = b_cPb$ or

$$\begin{aligned} E &= \gamma\left(E' + p'^1 w\right), \\ p^1 &= \gamma\left(p'^1 + \frac{E'}{c^2}w\right), \\ p^2 &= p'^2, \quad p^3 = p'^3. \end{aligned}$$

Under a general pure Lorentz transformation, one has $P' = bPb_c$ with $b = \cosh\frac{\varphi}{2} - i\frac{\mathbf{v}}{v}\sinh\frac{\varphi}{2}$.

6.3.2 Four-force

Let $P = j\frac{E}{c} + k\mathbf{p}$ be the four-momentum vector of the particle; the four-force vector is defined by

$$\begin{aligned} F &= \frac{dP}{d\tau} = \frac{d\left(m_0V\right)}{d\tau} \\ &= m_0A + \frac{dm_0}{d\tau}V \end{aligned}$$

with $V = j\gamma c + k\gamma\mathbf{v}$ [42, pp. 123-124]. If the four-force conserves m_0, then $\frac{dm_0}{d\tau} = 0$ and

$$\begin{aligned} F = m_0A &= \gamma\frac{d\left(jm_0\gamma c + k\gamma m_0\mathbf{v}\right)}{dt} \\ &= \gamma\left(v\right)\left(j\frac{dE}{cdt} + k\mathbf{f}\right) \end{aligned}$$

with, by definition $\mathbf{f} = \frac{d\mathbf{p}}{dt} = \frac{d(\gamma m_0 \mathbf{v})}{dt}$. Furthermore,

$$F \cdot V = \left(m_0 A + \frac{dm_0}{d\tau} V \right) \cdot V = \frac{dm_0}{d\tau} c^2$$
$$= \gamma^2 (v) \left(\frac{dE}{dt} - \mathbf{f} \cdot \mathbf{v} \right).$$

If $\frac{dm_0}{d\tau} = 0 = F \cdot V$, then $\frac{dE}{dt} = \mathbf{f} \cdot \mathbf{v}$ and

$$F = \gamma (v) \left[j \frac{(\mathbf{f} \cdot \mathbf{v})}{c} + k\mathbf{f} \right].$$

Since

$$\mathbf{f} = \frac{d(\gamma m_0 \mathbf{v})}{dt} = \gamma m_0 \mathbf{a} + \frac{d(\gamma m_0)}{dt} \mathbf{v}$$
$$= \gamma m_0 \mathbf{a} + \frac{dE}{c^2 dt} \mathbf{v} = \gamma m_0 \mathbf{a} + \frac{(\mathbf{f} \cdot \mathbf{v})}{c^2} \mathbf{v},$$

one infers that $\mathbf{a}$ is coplanar to $\mathbf{f}$ and $\mathbf{v}$ but is not in general parallel to $\mathbf{f}$. If $\frac{dm_0}{d\tau} = 0$, the force $\mathbf{f}$ satisfies the relation [42, p. 125]

$$\gamma (v) \mathbf{f} = m_0 \frac{d^2 \mathbf{x}}{dt^2} \tag{6.18}$$

which can be established as

$$F = m_0 A = m_0 \frac{d^2 X}{d\tau^2} \tag{6.19}$$
$$= \gamma (v) \left(j \frac{dE}{cdt} + k\mathbf{f} \right) \tag{6.20}$$
$$= m_0 \left[\frac{d^2 (jct)}{d\tau^2} + k \frac{d^2 \mathbf{x}}{dt^2} \right], \tag{6.21}$$

hence the relation (6.18) by comparing equations (6.20) and (6.21). Under a special Lorentz transformation, the four-force $F = \gamma (v) \left(j \frac{Q}{c} + k\mathbf{f} \right)$ with $Q = \frac{dE}{dt}$, transforms as $F' = bFb_c$ with $b = \left(\cosh \frac{\varphi}{2} - iI \sinh \frac{\varphi}{2} \right)$, $\tanh \varphi = \frac{w}{c}$ et $\cosh \varphi = \gamma = \frac{1}{\sqrt{1 - \frac{w^2}{c^2}}}$ or explicitly

$$F'^0 = \gamma (w) \left(F^0 - F^1 \frac{w}{c} \right),$$
$$F'^1 = \gamma (w) \left(F^1 - F^0 \frac{w}{c} \right),$$
$$F'^2 = F^2, \quad F'^3 = F^3,$$

and equivalently

$$\gamma(v')\,Q' = \gamma(w)\,\gamma(v)\left[Q - f^1 w\right], \tag{6.22}$$

$$\gamma(v')\,f'^1 = \gamma(w)\,\gamma(v)\left[f^1 - \frac{Qw}{c^2}\right], \tag{6.23}$$

$$\gamma(v')\,f'^2 = \gamma(w)\,f^2, \tag{6.24}$$

$$\gamma(v')\,f'^3 = \gamma(w)\,f^3. \tag{6.25}$$

Furthermore, one has the relation

$$\frac{\gamma(v')}{\gamma(v)} = \gamma(w)\left[1 - \frac{v^1 w}{c^2}\right], \tag{6.26}$$

which can be established as follows ([42, p. 69]); form the invariant

$$ds^2 = dt^2\left(c^2 - v^2\right) = dt'^2\left(c^2 - v'^2\right)$$

and the relation

$$\begin{aligned} cdt' &= \gamma(w)\left(cdt - dx\frac{w}{c}\right) \\ &= \gamma(w)\,cdt\left(1 - \frac{v^1 w}{c^2}\right). \end{aligned}$$

Then one obtains

$$\begin{aligned} dt^2\left(c^2 - v^2\right) &= dt^2\gamma^2(w)\left[1 - \frac{v^1 w}{c^2}\right]^2\left(c^2 - v'^2\right), \\ \left(c^2 - v'^2\right) &= \frac{\left(c^2 - v^2\right)\left(1 - \frac{w^2}{c^2}\right)}{\left(1 - \frac{v^1 w}{c^2}\right)^2}, \\ \frac{1}{\gamma^2(v')} &= \frac{1}{\gamma^2(v)}\frac{1}{\gamma^2(w)}\frac{1}{\left(1 - \frac{v^1 w}{c^2}\right)^2}, \end{aligned}$$

hence, the relation (6.26). Reciprocally, one has

$$\frac{\gamma(v)}{\gamma(v')} = \gamma(w)\left[1 + \frac{v'^1 w}{c^2}\right].$$

Finally, equations (6.22), (6.23), (6.24), (6.25) can be written [42, p. 124]

$$Q' = \frac{Q - f^1 w}{1 - \frac{v^1 w}{c^2}},$$

$$f'^1 = \frac{f^1 - \frac{wQ}{c^2}}{1 - \frac{v^1 w}{c^2}},$$

$$f'^2 = \frac{f^2}{\gamma(v)\left(1 - \frac{v^1 w}{c^2}\right)},$$

$$f'^3 = \frac{f^3}{\gamma(v)\left(1 - \frac{v^1 w}{c^2}\right)}.$$

6.4 Exercises

E6-1 Two particles A and B move towards the origin O in opposite directions (on the Ox axis) with a uniform velocity $0,8\,c$. Determine the relative velocity of B with respect to A for an observer at rest relatively to A.

E6-2 Consider an isolated set of particles without interaction in a reference frame K at rest. Determine the total relativistic angular momentum

$$L = \sum X_i \wedge P_i,$$

$$X_i = jct + kIx_i + kJy_i + kKz_i,$$

$$P_i = j\frac{E_i}{c} + kIp_{xi} + kJp_{yi} + kKp_{zi}.$$

Define the center of energy of the set. Show that it moves with a constant velocity.

E6-3 Consider a hyperbolic rectilinear motion of a particle whose acceleration is constant in the proper reference frame, at any instant. The particle being at rest at the origin of the axes and of the time, determine the four-vector $X = jct + k\mathbf{x}$ as a function of the parameter $\varphi = \frac{g\tau}{c}$ where τ is the proper time of the particle. Show that one obtains the classical results of a uniformly accelerated motion for small velocities $\varphi \ll 1$.

Chapter 7

Classical electromagnetism

Classical electromagnetism is treated within the Clifford algebra $\mathbb{H} \otimes \mathbb{H}$ over $\mathbb{R}$. This chapter develops Maxwell's equations, electromagnetic waves and relativistic optics.

7.1 Electromagnetic quantities

7.1.1 Four-current density and four-potential

Four-current density

Let ρ_0 be the charge density in the proper frame and $V = j\gamma c + k\gamma \mathbf{v}$ the four-velocity; the four-current density is defined by

$$\begin{aligned} C &= \rho_0 V = j\gamma\rho_0 c + k\gamma\rho_0 \mathbf{v} \\ &= j\rho c + k\mathbf{j} \end{aligned}$$

with $\rho = \gamma\rho_0$ and $\mathbf{j} = \rho\mathbf{v}$. One obtains the relativistic invariant

$$\begin{aligned} CC_c &= \rho^2 c^2 - \rho^2 (\mathbf{v})^2 = \rho^2 c^2 \left(1 - \frac{v^2}{c^2}\right) \\ &= c^2 \frac{\rho^2}{\gamma^2} = c^2 \rho_0^2 \end{aligned}$$

with $\gamma = \frac{1}{\sqrt{1-\frac{v^2}{c^2}}}$. Under a pure special Lorentz transformation, the four-current density transforms as $C' = bCb_c$ with $b = \cosh\frac{\varphi}{2} - iI\sinh\frac{\varphi}{2}$, $\tanh\varphi = \frac{w}{c} = \beta$ and $\cosh\varphi = \gamma = \frac{1}{\sqrt{1-\beta^2}}$; explicitly, one has

$$\begin{aligned} \rho' c &= \gamma\left(\rho c - j^1 \beta\right), \\ j'^1 &= \gamma\left(j^1 - \rho c \beta\right), \\ j'^2 &= j^2, \quad j'^3 = j^3. \end{aligned}$$

Consider a trivolume $OM(\alpha,\beta,\gamma) = jx^0(\alpha,\beta,\gamma) + k\mathbf{x}(\alpha,\beta,\gamma)$ depending on three parameters α,β,γ. The infinitesimal hyperplane dT is expressed by (4.17)

$$\begin{aligned} dT &= \left(\frac{\partial OM}{\partial \alpha} \wedge \frac{\partial OM}{\partial \beta} \wedge \frac{\partial OM}{\partial \gamma}\right) d\alpha d\beta d\gamma \\ &= \left(\begin{array}{l} kdx^1dx^3dx^2 + jIdx^2dx^3dx^0 \\ +jJdx^3dx^1dx^0 + jKdx^1dx^2dx^0 \end{array}\right). \end{aligned}$$

The dual dT^* of dT is given by

$$\begin{aligned} dT^* &= idT \\ &= \left(\begin{array}{l} jdx^1dx^2dx^3 + kIdx^2dx^3dx^0 \\ +kJdx^3dx^1dx^0 + kKdx^1dx^2dx^0 \end{array}\right) \end{aligned}$$

and is a four-vector orthogonal to dT. The relativistic invariant $C \cdot dT^*$ is given by

$$C \cdot dT^* = \left(\begin{array}{l} \rho c dx^1dx^2dx^3 - \rho v^1dx^2dx^3dx^0 \\ -\rho v^2dx^3dx^1dx^0 - \rho v^3dx^1dx^2dx^0 \end{array}\right);$$

in the proper frame $C \cdot dT^* = \rho_0 c dx^1dx^2dx^3 = cdq$ where dq is the electric charge contained in dT. The electric charge Q contained in the hyperplane T is given by

$$Q = \frac{1}{c}\iiint_T C \cdot dT^*$$

for any Galilean frame. Furthermore, if one integrates over a closed four-volume τ, one obtains

$$\oint_T C \cdot dT^* = \oint_\tau (\nabla \cdot C)\, dx^0dx^1dx^2dx^3$$

with the relation $\nabla \cdot C = \frac{\partial \rho c}{\partial x^0} + \frac{\partial \rho v^1}{\partial x^1} + \frac{\partial \rho v^2}{\partial x^2} + \frac{\partial \rho v^3}{\partial x^3} = 0$ expressing the conservation of electric charge. If the four-volume τ is limited by the hypersurfaces $H_1\,(t_1 = \text{const.})$, $H_2\,(t_2 = \text{const.})$ and the lateral hypersurface H_3 at infinity, one has (Figure 7.1)

$$\int_{H_1} C \cdot dT^* + \int_{H_2} C \cdot dT^* + \int_{H_3} C \cdot dT^* = 0;$$

furthermore, the integral over H_3 is nil in the absence of electric charges at infinity, hence

$$Q_2 = \int_{H_2} C \cdot dT^* = -\int_{H_1} C \cdot dT^* = Q_1$$

which expresses the conservation of electric charge.

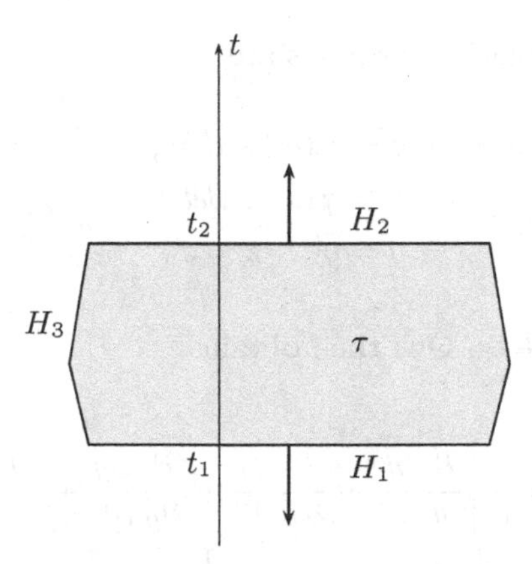

Figure 7.1: Conservation of electric charge: H_1 and H_2 are hyperplanes (at t constant) and H_3 is a lateral hypersurface at infinity.

Four-potential vector

Let V be the scalar potential and $\mathbf{A}$ the potential vector, the four-potential vector is defined by $A = j\frac{V}{c} + k\mathbf{A}$ yielding the relativistic invariant $AA_c = \frac{V^2}{c^2} - (\mathbf{A})^2$. Under a special Lorentz transformation, one has $A' = bAb_c$ with $b = \cosh\frac{\varphi}{2} - iI\sinh\frac{\varphi}{2}$, $\tanh\varphi = \frac{w}{c} = \beta$ and $\cosh\varphi = \gamma = \frac{1}{\sqrt{1-\beta^2}}$, hence

$$\begin{aligned} \frac{V'}{c} &= \gamma\left(\frac{V}{c} - A^1\beta\right), \\ A'^1 &= \gamma\left(A^1 - \frac{V}{c}\beta\right), \\ A'^2 &= A^2, \quad A'^3 = A^3. \end{aligned}$$

7.1.2 Electromagnetic field bivector

Four-nabla operator

The four-nabla operator

$$\nabla = j\frac{\partial}{c\partial t} - kI\frac{\partial}{\partial x^1} - kJ\frac{\partial}{\partial x^2} - kK\frac{\partial}{\partial x^3}$$

transforms under a general Lorentz transformation as a four-vector, i.e., $\nabla' = a\nabla a_c$. Let us demonstrate it explicitly in the case of a pure special Lorentz transformation. The transformation formulas are

$$\begin{aligned}
ct &= \gamma\left(ct + x'\beta\right),\\
x &= \gamma\left(x' + \beta ct'\right),\\
y &= y', \quad z = z',
\end{aligned}$$

with $\beta = \frac{v}{c}$ and $\gamma = \frac{1}{\sqrt{1-\frac{v^2}{c^2}}}$. One then obtains

$$\begin{aligned}
\frac{\partial}{c\partial t'} &= \frac{1}{c}\left[\frac{\partial}{\partial t}\frac{\partial t}{\partial t'} + \frac{\partial}{\partial x}\frac{\partial x}{\partial t'} + \frac{\partial}{\partial y}\frac{\partial y}{\partial t'} + \frac{\partial}{\partial z}\frac{\partial z}{\partial t'}\right]\\
&= \gamma\left[\frac{\partial}{c\partial t} - \left(-\frac{\partial}{\partial x}\right)\beta\right]
\end{aligned}$$

with $\frac{\partial t}{\partial t'} = \gamma$, $\frac{\partial x}{\partial t'} = \gamma\beta c$, $\frac{\partial y}{\partial t'} = \frac{\partial z}{\partial t'} = 0$. Furthermore,

$$\begin{aligned}
-\frac{\partial}{\partial x'} &= -\left[\frac{\partial}{\partial x}\frac{\partial x}{\partial x'} + \frac{\partial}{\partial t}\frac{\partial t}{\partial x'}\right]\\
&= \gamma\left[-\frac{\partial}{\partial x} - \beta\frac{\partial}{c\partial t}\right]
\end{aligned}$$

with $\frac{\partial x}{\partial x'} = \gamma$, $\frac{\partial t}{\partial x'} = \gamma\frac{\beta}{c}$. Finally, one has

$$\begin{aligned}
\frac{\partial}{c\partial t'} &= \gamma\left[\frac{\partial}{c\partial t} - \left(-\frac{\partial}{\partial x}\right)\beta\right],\\
-\frac{\partial}{\partial x'} &= \gamma\left[-\frac{\partial}{\partial x} - \beta\frac{\partial}{c\partial t}\right],\\
-\frac{\partial}{\partial y'} &= -\frac{\partial}{\partial y}, \qquad -\frac{\partial}{\partial z'} = -\frac{\partial}{\partial z},
\end{aligned}$$

which shows that the four-nabla operator transforms indeed as a four-vector, the demonstration being similar in the general case.

Electromagnetic field bivector

Let $A = j\frac{V}{c} + k\mathbf{A}$ be the four-potential vector, let us define the Clifford number

$$\begin{aligned}
F &= \nabla_c A = -\nabla A\\
&= (\nabla \cdot A) + (\nabla \wedge A)
\end{aligned}$$

and adopt the Lorentz gauge $(\nabla \cdot A) = \frac{\partial(\frac{V}{c})}{c\partial t} + \frac{\partial A^1}{\partial x^1} + \frac{\partial A^2}{\partial x^2} + \frac{\partial A^3}{\partial x^3} = 0$, which gives the electromagnetic field bivector

$$\begin{aligned} F &= \nabla \wedge A \\ &= -\,\mathbf{rot}\,\mathbf{A} + i\left(-\,\mathbf{grad}\,\frac{V}{c} - \frac{\partial \mathbf{A}}{c\partial t}\right) \\ &= -\mathbf{B} + i\frac{\mathbf{E}}{c} \end{aligned}$$

with the usual definitions of the magnetic induction $\mathbf{B}$ and the electric field $\mathbf{E}$,

$$\mathbf{B} = \mathbf{rot}\,\mathbf{A}, \qquad \mathbf{E} = -\,\mathbf{grad}\,V - \frac{\partial \mathbf{A}}{\partial t}. \tag{7.1}$$

Under a gauge transformation

$$A \to A' = A + \nabla f$$

where f is an arbitrary scalar function, one obtains the same electromagnetic field bivector

$$\begin{aligned} F' &= \nabla \wedge A' = \nabla \wedge A + (\nabla \wedge \nabla)\, f \\ &= \nabla \wedge A = F. \end{aligned}$$

The electromagnetic field bivector yields the relativistic invariant

$$\begin{aligned} F^2 &= F \cdot F + F \wedge F \\ &= \left[-\left(\mathbf{B}\right)^2 + \frac{\left(\mathbf{E}\right)^2}{c^2} + 2i\frac{\mathbf{E}\cdot\mathbf{B}}{c}\right]. \end{aligned}$$

Under a pure special Lorentz transformation , one has

$$F' = bFb_c$$

with $b = \cosh\frac{\varphi}{2} - iI\sinh\frac{\varphi}{2}$, $\tanh\varphi = \frac{v}{c} = \beta$ and $\cosh\varphi = \gamma = \frac{1}{\sqrt{1-\frac{v^2}{c^2}}}$, hence

$$\begin{aligned} F' &= \left[\begin{array}{l} -B^1 I - \left(B^2\cosh\varphi + \dfrac{E^3}{c}\sinh\varphi\right)J \\ +\left(-B^3\cosh\varphi + \dfrac{E^2}{c}\sinh\varphi\right)K \\ +\dfrac{E^1}{c}iI + \left(\dfrac{E^2}{c}\cosh\varphi - B^3\sinh\varphi\right)iJ \\ +\left(\dfrac{E^3}{c}\cosh\varphi + B^2\sinh\varphi\right)iK \end{array}\right] \\ &= -\mathbf{B}' + i\frac{\mathbf{E}'}{c} \end{aligned}$$

or

$$B'^1 = B^1, \qquad B'^2 = \gamma\left(B^2 + \frac{E^3}{c}\beta\right), \qquad B'^3 = \gamma\left(B^3 - \frac{E^2}{c}\beta\right),$$
$$E'^1 = E^1, \qquad E'^2 = \gamma\left(E^2 - B^3\beta c\right), \qquad E'^3 = \gamma\left(E^3 + B^2\beta c\right).$$

Under a general pure Lorentz transformation, one has $F' = bFb_c$ with $b = \cosh\frac{\varphi}{2} - i\frac{\mathbf{v}}{v}\sinh\frac{\varphi}{2}$, $\tanh\varphi = \frac{v}{c} = \beta$ and $\cosh\varphi = \gamma = \frac{1}{\sqrt{1-\frac{v^2}{c^2}}}$; hence the formulas [40, p. 191]

$$\mathbf{B}' = \gamma\left[\mathbf{B} + \frac{\mathbf{v}}{v^2}(\mathbf{v}\cdot\mathbf{B})\left(\frac{1}{\gamma} - 1\right) - \frac{1}{c^2}(\mathbf{v}\times\mathbf{E})\right],$$
$$\mathbf{E}' = \gamma\left[\mathbf{E} + \frac{\mathbf{v}}{v^2}(\mathbf{v}\cdot\mathbf{E})\left(\frac{1}{\gamma} - 1\right) + (\mathbf{v}\times\mathbf{B})\right].$$

7.2 Maxwell's equations

7.2.1 Differential formulation

In vacuum

With the Lorentz gauge $\nabla\cdot A = 0$, one has

$$\begin{aligned} F &= \nabla\wedge A = -\mathbf{B} + i\frac{\mathbf{E}}{c}, \\ \nabla\wedge F &= \nabla\wedge\nabla\wedge A = 0, \\ &= k(-\operatorname{div}\mathbf{B}) + j\left(-\frac{\partial\mathbf{B}}{c\partial t} - \mathbf{rot}\,\frac{\mathbf{E}}{c}\right); \end{aligned}$$

hence, one obtains two of the Maxwell's equations

$$\operatorname{div}\mathbf{B} = 0, \qquad \mathbf{rot}\,\mathbf{E} = -\frac{\partial\mathbf{B}}{c\partial t}.$$

The two other equations are given by

$$\begin{aligned} \nabla\cdot F &= \mu_0 C \\ &= j\operatorname{div}\frac{\mathbf{E}}{c} + k\left(\mathbf{rot}\,\mathbf{B} - \frac{1}{c^2}\frac{\partial\mathbf{E}}{\partial t}\right) \end{aligned}$$

with the four-current density $C = j\rho c + k\rho\mathbf{v}$; one obtains with $\varepsilon_0\mu_0 c^2 = 1$,

$$\begin{aligned} \operatorname{div}\mathbf{E} &= \mu_0\rho c^2 = \frac{\rho}{\varepsilon_0}, \\ \mathbf{rot}\,\mathbf{B} &= \frac{1}{c^2}\frac{\partial\mathbf{E}}{\partial t} + \mu_0\mathbf{j}. \end{aligned}$$

The complete set of Maxwell's equations reads

$$\begin{aligned}
\nabla F &= \nabla\nabla_c A = \nabla \cdot F + \nabla \wedge F \\
&= \begin{bmatrix} j \operatorname{div} \dfrac{\mathbf{E}}{c} + k\left(\mathbf{rot}\,\mathbf{B} - \dfrac{1}{c^2}\dfrac{\partial \mathbf{E}}{\partial t}\right) \\ +k\,(-\operatorname{div}\mathbf{B}) + j\left(-\dfrac{\partial \mathbf{B}}{c\partial t} - \mathbf{rot}\,\dfrac{\mathbf{E}}{c}\right) \end{bmatrix} \\
&= \Box A = \mu_0 C
\end{aligned}$$

with the d'Alembertian operator $\Box = \nabla\nabla_c = \frac{\partial^2}{c^2\partial t^2} - \triangle$ and $A = j\frac{V}{c} + k\mathbf{A}$ the four-potential vector. Hence, the equations

$$\begin{aligned}
\Box\frac{V}{c} &= \rho\mu_0 c^2 = \frac{\rho}{\varepsilon_0}, \\
\Box\mathbf{A} &= \mu_0 \mathbf{j},
\end{aligned}$$

or equivalently

$$\begin{aligned}
&\operatorname{div}\mathbf{B} = 0, \qquad \operatorname{div}\mathbf{E} = \frac{\rho}{\varepsilon_0}, \\
&\mathbf{rot}\,\mathbf{E} = -\frac{\partial \mathbf{B}}{\partial t}, \\
&\mathbf{rot}\,\mathbf{B} = \mu_0\left[\mathbf{j} + \frac{\partial \varepsilon_0 \mathbf{E}}{\partial t}\right].
\end{aligned}$$

If one does not adopt the Lorentz gauge, one obtains

$$\begin{aligned}
F &= \nabla_c A = -\nabla A \\
&= \left[\left(\operatorname{div}\mathbf{A} + \frac{\partial V}{c\partial t}\right) - \mathbf{B} + i\frac{\mathbf{E}}{c}\right]
\end{aligned}$$

which is not a bivector but an element of C^+. One obtains the same Maxwell's equations

$$\begin{aligned}
\nabla F &= \nabla\nabla_c A \\
&= \Box A = \mu_0 C.
\end{aligned}$$

Since $\nabla \cdot C = 0$ expresses the conservation of electric charge, one has

$$\begin{aligned}
\nabla \cdot (\Box A) &= \mu_0 \nabla \cdot C \\
&= \Box\,(\nabla \cdot A) = 0,
\end{aligned}$$

a condition which is less restrictive than the Lorentz gauge $\nabla \cdot A = 0$. The covariance of Maxwell's equations is manifest since under a general Lorentz transformation any Clifford number X transforms as $X' = aXa_c$ $(aa_c = 1)$ and thus

$$\begin{aligned}
\nabla' F' &= \mu_0 C' = a\nabla F a_c \\
&= a\mu_0 C a_c;
\end{aligned}$$

hence, $\nabla F = \mu_0 C$.

Perfect dielectric or magnetic medium

Consider the bivectors $F = -\mathbf{B} + i\frac{\mathbf{E}}{c}$, $G = -\mathbf{H} + ic\mathbf{D}$. Maxwell's equations are expressed by

$$\begin{aligned}\nabla \wedge F &= 0,\\ \nabla \cdot G &= C = j\rho c + k\rho\mathbf{v}\\ &= j\operatorname{div}(c\mathbf{D}) + k\left(\mathbf{rot}\,\mathbf{H} - \frac{\partial \mathbf{D}}{\partial t}\right)\end{aligned}$$

or

$$\operatorname{div}\mathbf{B} = 0, \qquad \operatorname{div}\mathbf{D} = \rho,$$

$$\mathbf{rot}\,\mathbf{E} = -\frac{\partial \mathbf{B}}{\partial t}, \qquad \mathbf{rot}\,\mathbf{H} = \mathbf{j} + \frac{\partial \mathbf{D}}{\partial t}.$$

The bivector G yields the relativistic invariant

$$\begin{aligned}G^2 &= G \cdot G + G \wedge G\\ &= -\mathbf{H}^2 + c^2\mathbf{D}^2 + 2ic\mathbf{D}\cdot\mathbf{H}.\end{aligned}$$

Under a pure special Lorentz transformation, G transforms as $G' = bGb_c$ with $b = \cosh\frac{\varphi}{2} - iI\sinh\frac{\varphi}{2}$, $\tanh\varphi = \frac{v}{c} = \beta$ and $\cosh\varphi = \gamma = \frac{1}{\sqrt{1-\frac{v^2}{c^2}}}$, hence

$$\begin{aligned}H'_x &= H_x,\\ H'_y &= \gamma\left(H_y + vD_z\right),\\ H'_z &= \gamma\left(H_z - vD_y\right),\\ D'_x &= D_x,\\ D'_y &= \gamma\left(D_y - v\frac{H_z}{c^2}\right),\\ D'_z &= \gamma\left(D_z + v\frac{H_y}{c^2}\right).\end{aligned}$$

Under a general pure Lorentz transformation, one has $G' = bGb_c$ with $b = \cosh\frac{\varphi}{2} - i\frac{\mathbf{v}}{v}\sinh\frac{\varphi}{2}$, $\tanh\varphi = \frac{v}{c} = \beta$ and $\cosh\varphi = \gamma = \frac{1}{\sqrt{1-\frac{v^2}{c^2}}}$; hence

$$\mathbf{H}' = \gamma\left[\mathbf{H} + \frac{\mathbf{v}}{v^2}(\mathbf{v}\cdot\mathbf{H})\left(\frac{1}{\gamma} - 1\right) - (\mathbf{v}\times\mathbf{D})\right], \tag{7.2}$$

$$\mathbf{D}' = \gamma\left[\mathbf{D} + \frac{\mathbf{v}}{v^2}(\mathbf{v}\cdot\mathbf{D})\left(\frac{1}{\gamma} - 1\right) + \left(\mathbf{v}\times\frac{\mathbf{H}}{c^2}\right)\right], \tag{7.3}$$

and the inverse formulas

$$\mathbf{H} = \gamma \left[\mathbf{H}' + \frac{\mathbf{v}}{v^2} (\mathbf{v} \cdot \mathbf{H}') \left(\frac{1}{\gamma} - 1 \right) + (\mathbf{v} \times \mathbf{D}') \right],$$

$$\mathbf{D} = \gamma \left[\mathbf{D}' + \frac{\mathbf{v}}{v^2} (\mathbf{v} \cdot \mathbf{D}') \left(\frac{1}{\gamma} - 1 \right) - \left(\mathbf{v} \times \frac{\mathbf{H}'}{c^2} \right) \right].$$

As an application, consider a perfect medium in the proper reference frame K' with $\mathbf{B}' = \mu_0 \mu_r \mathbf{H}'$, $\mathbf{D}' = \varepsilon_0 \varepsilon_r \mathbf{E}'$, moving with respect to a Galilean reference frame (at rest) K with a velocity $\mathbf{v}$. Using the transformation formulas (7.2), (7.3), one obtains

$$\left[\mathbf{B} + \frac{\mathbf{v}}{v^2} (\mathbf{v} \cdot \mathbf{B}) \left(\frac{1}{\gamma} - 1 \right) - \frac{1}{c^2} (\mathbf{v} \times \mathbf{E}) \right] \tag{7.4}$$

$$= \mu_0 \mu_r \left[\mathbf{H} + \frac{\mathbf{v}}{v^2} (\mathbf{v} \cdot \mathbf{H}) \left(\frac{1}{\gamma} - 1 \right) - (\mathbf{v} \times \mathbf{D}) \right], \tag{7.5}$$

$$\left[\mathbf{D} + \frac{\mathbf{v}}{v^2} (\mathbf{v} \cdot \mathbf{D}) \left(\frac{1}{\gamma} - 1 \right) + \left(\mathbf{v} \times \frac{\mathbf{H}}{c^2} \right) \right] \tag{7.6}$$

$$= \varepsilon_0 \varepsilon_r \left[\mathbf{E} + \frac{\mathbf{v}}{v^2} (\mathbf{v} \cdot \mathbf{E}) \left(\frac{1}{\gamma} - 1 \right) + (\mathbf{v} \times \mathbf{B}) \right]. \tag{7.7}$$

Furthermore, from equations (7.5), (7.7) one deduces the relations

$$\mathbf{v} \cdot \mathbf{B} = \mu_0 \mu_r \mathbf{v} \cdot \mathbf{H},$$

$$\mathbf{v} \cdot \mathbf{D} = \varepsilon_0 \varepsilon_r \mathbf{v} \cdot \mathbf{E}.$$

Equations (7.5), (7.7) then become

$$\mathbf{B} - \frac{1}{c^2} (\mathbf{v} \times \mathbf{E}) = \mu_0 \mu_r \left[\mathbf{H} - (\mathbf{v} \times \mathbf{D}) \right], \tag{7.8}$$

$$\mathbf{D} + \left(\mathbf{v} \times \frac{\mathbf{H}}{c^2} \right) = \varepsilon_0 \varepsilon_r \left[\mathbf{E} + (\mathbf{v} \times \mathbf{B}) \right]; \tag{7.9}$$

taking $\mathbf{H}$ from equation (7.8) and replacing it in equation (7.9) one finds [7, p. 239]

$$\mathbf{D} = \varepsilon_0 \varepsilon_r \mathbf{E} + \frac{\varepsilon_0 \left(\varepsilon_r - \frac{1}{\mu_r} \right)}{1 - \beta^2} \mathbf{v} \times \left(\mathbf{B} - \mathbf{v} \times \frac{\mathbf{E}}{c^2} \right).$$

Similarly, taking $\mathbf{D}$ from equation (7.9) and replacing it in equation (7.8) one obtains

$$\mathbf{H} = \frac{\mathbf{B}}{\mu_0 \mu_r} + \frac{\varepsilon_0 \left(\varepsilon_r - \frac{1}{\mu_r} \right)}{1 - \beta^2} \mathbf{v} \times (\mathbf{E} + \mathbf{v} \times \mathbf{B}).$$

One observes that $\mathbf{D}$ and $\mathbf{H}$ depend on $\mathbf{E}$ and $\mathbf{B}$ as well as on the velocity of the material.

Arbitrary dielectric or magnetic medium

In an arbitrary dielectric or magnetic medium, with a polarization density $\mathbf{P}$ and a magnetization density $\mathbf{M}$, one has the relations

$$\mathbf{D} = \varepsilon_0 \mathbf{E} + \mathbf{P},$$
$$\mathbf{B} = \mu_0 (\mathbf{H} + \mathbf{M})$$

which one can write in the form

$$F = \mu_0 (G + N)$$

with the bivectors $F = -\mathbf{B} + i\frac{\mathbf{E}}{c}$, $G = -\mathbf{H} + ic\mathbf{D}$ and $N = -\mathbf{M} - ic\mathbf{P}$. Maxwell's equation $\nabla \cdot G = C$ (with $C = j\rho c + k\rho\mathbf{v}$) then becomes

$$\nabla \cdot G = \nabla \cdot \left[\frac{F}{\mu_0} - N\right] = C$$

or

$$\nabla \cdot F = \mu_0 (C + \nabla \cdot N)$$

with

$$\nabla \cdot N = j \operatorname{div}(-c\mathbf{P}) + k\left(\mathbf{rot}\,\mathbf{M} + \frac{\partial \mathbf{P}}{\partial t}\right),$$
$$\nabla \cdot F = j \operatorname{div}\left(\frac{E}{c}\right) + k\left(\mathbf{rot}\,\mathbf{B} - \frac{1}{c^2}\frac{\partial \mathbf{E}}{\partial t}\right),$$

hence, the relations

$$\operatorname{div} \mathbf{E} = \frac{1}{c}(\rho - \operatorname{div} \mathbf{P}),$$
$$\mathbf{rot}\,\mathbf{B} = \mu_0 \left(\mathbf{j} + \frac{\partial \mathbf{P}}{\partial t} + \mathbf{rot}\,\mathbf{M}\right) + \varepsilon_0\mu_0 \frac{\partial \mathbf{E}}{\partial t}.$$

The entire set of Maxwell's equations is expressed by (with $\nabla \wedge F = 0$)

$$\nabla F = \nabla \cdot F + \nabla \wedge F = \Box A$$
$$= \mu_0 C + \mu_0 \nabla \cdot N$$

or

$$\Box V = \frac{\rho - \operatorname{div} P}{\varepsilon_0},$$
$$\Box \mathbf{A} = \mu_0 \left(\mathbf{j} + \frac{\partial \mathbf{P}}{\partial t} + \mathbf{rot}\,\mathbf{M}\right).$$

Consequently, one can replace the medium by a distribution, in vacuum, of charge density $\rho' = \operatorname{div} \mathbf{P}$ and a current density $\mathbf{j}' = \frac{\partial \mathbf{P}}{\partial t} + \mathbf{rot}\, \mathbf{M}$. Under a general pure Lorentz transformation, the bivector magnetization-polarization N transforms as $N' = bNb_c$ with $b = \cosh \frac{\varphi}{2} - i \frac{\mathbf{v}}{v} \sinh \frac{\varphi}{2}$, $\tanh \varphi = \frac{v}{c} = \beta$ and $\cosh \varphi = \gamma = \frac{1}{\sqrt{1-\frac{v^2}{c^2}}}$, hence

$$\mathbf{M}' = \gamma \left[\mathbf{M} + \frac{\mathbf{v}}{v^2} (\mathbf{v} \cdot \mathbf{M}) \left(\frac{1}{\gamma} - 1 \right) + (\mathbf{v} \times \mathbf{P}) \right],$$
$$\mathbf{P}' = \gamma \left[\mathbf{P} + \frac{\mathbf{v}}{v^2} (\mathbf{v} \cdot \mathbf{P}) \left(\frac{1}{\gamma} - 1 \right) - \left(\mathbf{v} \times \frac{\mathbf{M}}{c^2} \right) \right]$$

and the inverse formulas

$$\mathbf{M} = \gamma \left[\mathbf{M}' + \frac{\mathbf{v}}{v^2} (\mathbf{v} \cdot \mathbf{M}') \left(\frac{1}{\gamma} - 1 \right) - (\mathbf{v} \times \mathbf{P}') \right],$$
$$\mathbf{P} = \gamma \left[\mathbf{P}' + \frac{\mathbf{v}}{v^2} (\mathbf{v} \cdot \mathbf{P}') \left(\frac{1}{\gamma} - 1 \right) + \left(\mathbf{v} \times \frac{\mathbf{M}'}{c^2} \right) \right].$$

For low velocities ($\gamma \simeq 1$), one obtains

$$\mathbf{M} = \mathbf{M}' - \mathbf{v} \times \mathbf{P}',$$
$$\mathbf{P} = \mathbf{P}' + \mathbf{v} \times \frac{\mathbf{M}'}{c^2}.$$

If, in the proper reference frame K', one has $\mathbf{M}' = 0$ and $\mathbf{P}' \neq 0$, then in the reference frame at rest K one has $\mathbf{M} = -\mathbf{v} \times \mathbf{P}'$; if in K', $\mathbf{P}' = 0$, $\mathbf{M}' \neq 0$, then one obtains in K a polarization

$$\mathbf{P} = \mathbf{v} \times \frac{\mathbf{M}'}{c^2}.$$

7.2.2 Integral formulation

The relativistic integral formulation of Maxwell's equations in vacuum results from the general formulas valid for any bivector F (4.20), (4.21)

$$\oint F \cdot dS = \int (\nabla \wedge F) \cdot dT,$$
$$\oint F \wedge dS = - \int (\nabla \cdot F) \wedge dT,$$

with $dS = \mathbf{dS} + i\mathbf{dS}'$,

$$\mathbf{dS} = dydzI + dzdxJ + dxdyK,$$
$$\mathbf{dS}' = cdxdtI + cdydtJ + cdzdtK,$$
$$dT = -kdxdydz + j\,(dydzcdtI + dzdxcdtJ + dxdycdtK),$$

and where the integration is taken on a closed surface . Considering $F = -\mathbf{B} + i\frac{\mathbf{E}}{c}$ with $\nabla \wedge F = 0$ and $\nabla \cdot F = \mu_0 C$, one obtains [4, p. 227]

$$\begin{aligned}\oint F \cdot dS &= \oint \begin{pmatrix} B_x dydz + B_y dzdx + B_z dxdy \\ +E_x dxdt + E_y dydt + E_z dzdt \end{pmatrix} \\ &= 0, \\ \oint F \wedge dS &= \frac{i}{c} \int \begin{pmatrix} E_x dydz + E_y dzdx + E_z dxdy \\ -B_x dxcdt - B_y dycdt - B_z dzcdt \end{pmatrix} \\ &= i\mu_0 \int (\rho c dxdydz - j_x dydzdt - j_y dzdxcdt - j_z dxdycdt) .\end{aligned}$$

If one operates at t constant, one finds again the standard equations (in classical three-vector formulation)

$$\oint \mathbf{B} \cdot \mathbf{dS} = 0, \oint \mathbf{E} \cdot \mathbf{dS} = \frac{q}{\varepsilon_0}.$$

Furthermore, using the general formula (4.18)

$$\oint A \cdot dl = -\int (\nabla \wedge A) \cdot dS$$

one obtains for the four-potential vector A (with the Lorentz gauge $\nabla \cdot A = 0$)

$$\oint A \cdot dl = -\int F \cdot dS$$

or [4, p. 231]

$$\oint \mathbf{A} \cdot \mathbf{dx} - V dt = \int \begin{pmatrix} B_x dydz + B_y dzdx + B_z dxdy \\ +E_x dxdt + E_y dydt + E_z dzdt \end{pmatrix}.$$

At t constant, one obtains the classical vector formulation of Stokes' theorem

$$\oint \mathbf{A} \cdot \mathbf{dl} = \int \mathbf{rot\,A} \cdot \mathbf{dS} = \int \mathbf{B} \cdot \mathbf{dS}.$$

7.2.3 Lorentz force

The four-force density (per unit volume of the laboratory frame) is expressed with $F = -\mathbf{B} + i\frac{\mathbf{E}}{c}$ and $C = j\rho c + k\rho\mathbf{v}$ by

$$\begin{aligned} f &= F \cdot C \\ &= j(\mathbf{j} \cdot \mathbf{E}) + k(\rho\mathbf{E} + \mathbf{j} \times \mathbf{B}) ;\end{aligned}$$

f can also be expressed in the form (with $\nabla F = \mu_0 C$ and $\nabla = \partial_\alpha e^\alpha$)

$$\begin{aligned} f = \frac{FC - CF}{2} &= \frac{1}{2\mu_0}\left[F(\nabla F) - (\nabla F)F\right] \\ &= \frac{1}{2\mu_0}\left[F(\partial_\alpha e^\alpha F) - (\partial_\alpha e^\alpha F)F\right] \\ &= \frac{1}{2\mu_0}\left[\partial_\alpha(Fe^\alpha F) - (\partial_\alpha F)e^\alpha F - (\partial_\alpha e^\alpha F)F\right]. \end{aligned}$$

Since

$$\begin{aligned} \nabla \wedge F &= \frac{\nabla F + F\nabla}{2} \\ &= \frac{1}{2}\left[\partial_\alpha(e^\alpha F) + (\partial_\alpha F)e^\alpha\right] = 0 \end{aligned}$$

it follows that

$$f = \frac{1}{2\mu_0}\partial_\alpha(Fe^\alpha F) = -\partial_\alpha t^\alpha$$

with

$$t^\alpha = \frac{-Fe^\alpha F}{2\mu_0} = T^{\alpha\beta}e_\beta$$

where $T^{\alpha\beta}$ is the energy-momentum tensor; t^α is a four-vector since $(t^\alpha)_c = -t^\alpha$ (with $F_c = -F$). One obtains for t^α,

$$t^0 = j\frac{(\mathbf{B}\cdot\mathbf{H} + \mathbf{E}\cdot\mathbf{D})}{2} + k\frac{\mathbf{E}\times\mathbf{H}}{c},$$

$$t^1 = j\frac{(\mathbf{E}\times\mathbf{H})_1}{c} + \frac{k}{\mu_0}\left[\begin{array}{l} \left(\frac{\mathbf{B}^2 + \frac{\mathbf{E}^2}{c^2}}{2} - B_1^2 - \frac{E_1^2}{c^2}\right)I \\ -\left(B_1B_2 + \frac{E_1E_2}{c^2}\right)J - \left(B_1B_3 + \frac{E_1E_3}{c^2}\right)K \end{array}\right],$$

$$t^2 = j\frac{(\mathbf{E}\times\mathbf{H})_2}{c} + \frac{k}{\mu_0}\left[\begin{array}{l} -\left(B_1B_2 + \frac{E_1E_2}{c^2}\right)I \\ \left(\frac{\mathbf{B}^2 + \frac{\mathbf{E}^2}{c^2}}{2} - B_2^2 - \frac{E_2^2}{c^2}\right)J - \left(B_2B_3 + \frac{E_2E_3}{c^2}\right)K \end{array}\right],$$

$$t^3 = j\frac{(\mathbf{E}\times\mathbf{H})_3}{c} + \frac{k}{\mu_0}\left[\begin{array}{l} -\left(B_1B_3 + \frac{E_1E_3}{c^2}\right)I \\ -\left(B_2B_3 + \frac{E_2E_3}{c^2}\right)J + \left(\frac{\mathbf{B}^2 + \frac{\mathbf{E}^2}{c^2}}{2} - B_3^2 - \frac{E_3^2}{c^2}\right)K \end{array}\right].$$

The four-vectors t^α constitute the energy-current density ($\alpha = 0$) and the momentum-current density along the three axes (x, y, z); the α component of the four-momentum of the electromagnetic field of a trivolume T is given by

$$P^\alpha = \int t^\alpha \cdot (dT^*).$$

The symmetry of the energy-momentum tensor $T^{\alpha\beta}$ of the electromagnetic field can be verified directly on the components of t^α

$$\begin{aligned} t^\alpha \cdot e^\beta &= t^\beta \cdot e^\alpha \\ &= T^{\alpha\gamma} e_\gamma \cdot e^\beta = T^{\alpha\beta} = T^{\beta\alpha}; \end{aligned}$$

it is sufficient to verify the equality

$$(Fe^\alpha F) \cdot e^\beta = e^\alpha \cdot \left(Fe^\beta F\right), \tag{7.10}$$

$$Fa = aF \tag{7.11}$$

with $a = e^\alpha F e^\beta - e^\beta F e^\alpha \in C^+$; since $a = a_c$, a is constituted by a scalar and a pseudoscalar, and thus commutes with F, which demonstrates the equality (7.10).

7.3 Electromagnetic waves

7.3.1 Electromagnetic waves in vacuum

Starting from Maxwell's equations $\nabla F = \mu_0 C$ with $F = -\mathbf{B} + i\frac{\mathbf{E}}{c}$, $C = j\rho c + k\mathbf{j}$, one obtains in the absence of electric charges and currents

$$\nabla_c \nabla F = \Box F = 0$$

where $\Box = \nabla\nabla_c = \frac{\partial^2}{c^2 \partial t^2} - \Delta$ is the d'Alembertian operator; hence

$$\Box \mathbf{B} = 0, \qquad \Box \mathbf{E} = 0.$$

A sinusoidal plane wave rectilinearly polarized is, in complex notation, of the type

$$\underline{F} = F_m e^{i'(\omega t - \mathbf{k}\cdot\mathbf{r})} = -\underline{\mathbf{B}} + i\frac{\underline{\mathbf{E}}}{c}$$

where $F_m = -\mathbf{B}_m + i\frac{\mathbf{E}_m}{c}$ is a constant bivector ($\mathbf{B}_m, \mathbf{E}_m$ real). One has, in a vacuum,

$$\begin{aligned} \nabla \underline{F} &= i'\left(j\frac{\omega}{c} + k\mathbf{k}\right)\underline{F} \\ &= i' \left\{ \begin{array}{l} j\left(\frac{-\mathbf{k}\cdot\underline{\mathbf{E}}}{c}\right) + k\left(\mathbf{k}\cdot\underline{\mathbf{B}}\right) \\ +j\left(\mathbf{k}\times\underline{\mathbf{E}} - \underline{\mathbf{B}}\omega\right) + k\left(-\mathbf{k}\times\underline{\mathbf{B}} - \frac{\omega\underline{\mathbf{E}}}{c^2}\right) \end{array} \right\} \\ &= 0; \end{aligned}$$

hence, the relations characterizing the plane wave

$$\mathbf{k}\cdot\mathbf{E}=0, \tag{7.12}$$
$$\mathbf{k}\cdot\mathbf{B}=0, \tag{7.13}$$
$$\mathbf{B}=\frac{\mathbf{k}\times\mathbf{E}}{\omega}, \tag{7.14}$$
$$\mathbf{E}=\frac{-\left(\mathbf{k}\times\mathbf{B}\right)c^2}{\omega}. \tag{7.15}$$

The relation

$$\nabla_c\nabla\underline{F}=\left(\frac{\omega}{c}j+k\mathbf{k}\right)^2\underline{F}=\left(-\frac{\omega^2}{c^2}+\mathbf{k}\cdot\mathbf{k}\right)\underline{F}=0$$

gives the norm of $\mathbf{k}$, $|\mathbf{k}|=\frac{\omega}{c}$ and the phase velocity $v=\frac{\omega}{|\mathbf{k}|}=c$.

For a plane wave polarized elliptically, one just needs to take F_m complex.

7.3.2 Electromagnetic waves in a conductor

Consider a conductor having a permittivity $\epsilon=\varepsilon_0$, a permeability $\mu=\mu_0$ and a conductivity γ ($\mathbf{j}=\gamma\mathbf{E}$). Maxwell's equations are expressed by $\nabla F=\mu_0 C$, hence

$$\nabla_c\nabla F=\Box F=\mu_0\nabla_c C \tag{7.16}$$
$$=\mu_0\left(\nabla\cdot C+\nabla\wedge C\right)=\mu_0\left(\nabla\wedge C\right) \tag{7.17}$$
$$=\mu_0\left[-\mathbf{rot}\,\mathbf{j}+i\left(-\frac{\partial\mathbf{j}}{c\partial t}-(c)\,\mathbf{grad}\,\rho\right)\right] \tag{7.18}$$

with the conservation of electric charge ($\nabla\cdot C=0$). Using the relations $\mathbf{j}=\gamma\mathbf{E}$ and $\nabla\wedge F=0$, equation (7.18) becomes

$$\Box F+\gamma\mu_0\frac{\partial F}{\partial t}=-i(\mu_0 c)\,\mathbf{grad}\,\rho.$$

In the absence of an electric charge density ($\rho=0$), one has

$$\Box F+\gamma\mu_0\frac{\partial F}{\partial t}=0. \tag{7.19}$$

Consider a sinusoidal plane wave rectilinearly polarized $\underline{F}=F_m e^{i'\,t(\omega t-\mathbf{k}\cdot\mathbf{r})}$ with the constant bivector $F_m=-\mathbf{B}_m+i\frac{\mathbf{E}_m}{c}$. Equation (7.19) gives with

$$\frac{\partial\underline{F}}{\partial t}=i'\omega\underline{F},\qquad \Box F=\left[\left(-\frac{\omega^2}{c^2}+|\mathbf{k}|^2\right)\underline{F}\right]$$

a complex expression of $\mathbf{k}$

$$|\mathbf{k}|^2=\frac{\omega^2}{c^2}-i'\omega\gamma\mu_0.$$

The structure of the plane wave results from equation $\nabla\underline{F}=\mu_0\underline{C}$ with $\underline{C}=C_m e^{i'(\omega t-\mathbf{k}\cdot\mathbf{r})}$ and yields the relations (7.12), (7.13), (7.14), (7.15) with $\mathbf{k}$ complex.

7.3.3 Electromagnetic waves in a perfect medium

Wave equation

In a perfect medium, one has the relations

$$\nabla \wedge F = 0, \qquad \nabla \cdot G = C$$

with $F = -\mathbf{B} + i\frac{\mathbf{E}}{c}$, $G = -\mathbf{H} + ic\mathbf{D}$, $C = j\rho c + k\mathbf{j}$. Hence

$$\nabla \wedge (\nabla \cdot G) = \nabla \wedge C \tag{7.20}$$

with

$$\nabla \wedge (\nabla \cdot G) = \left\{ \begin{array}{l} \left[\mathbf{rot}\,\frac{\partial \mathbf{D}}{\partial t} - \mathbf{rot}\,(\mathbf{rot}\,\mathbf{H})\right] \\ +i\left[\frac{\partial^2 \mathbf{D}}{c\partial t^2} - (c)\,\mathbf{grad}\,(\operatorname{div}\mathbf{D}) - \frac{\partial\,\mathbf{rot}\,\mathbf{H}}{c\partial t}\right] \end{array} \right\},$$

$$\nabla \wedge C = -\,\mathbf{rot}\,j + i\left(-\frac{\partial \mathbf{j}}{c\partial t} - (c)\,\mathbf{grad}\,\rho\right).$$

Equation (7.20) can be written in the form

$$\triangle G - \varepsilon\mu\frac{\partial^2 G}{\partial t^2} = \gamma\mu\frac{\partial G}{\partial t} + \mathbf{grad}\,(\operatorname{div}\mathbf{D})\,.$$

In a nonconducting medium ($\gamma = 0$) and in the absence of an electric charge density, one obtains the wave equation

$$\triangle G - \varepsilon\mu\frac{\partial^2 G}{\partial t^2} = 0. \tag{7.21}$$

Introducing the operator $\nabla' = j\frac{\partial}{v\partial t} - k\boldsymbol{\nabla}$, $G' = -H + vD$, $F' = -\mathbf{B} + i\frac{\mathbf{E}}{v} = \mu G'$, $C' = j\rho v + k\mathbf{j}$ and the constant $v = \frac{1}{\sqrt{\varepsilon\mu}}$, Maxwell's equations are expressed by [5]

$$\begin{aligned} \nabla' \wedge G' &= 0, \\ \nabla' \cdot G' &= C' \end{aligned}$$

or $\nabla' G' = C'$. Hence, the wave equation

$$\begin{aligned} \nabla'_c \nabla' G' &= \nabla'_c C' \\ &= \Box G'. \end{aligned}$$

In a nonconducting medium ($\gamma = 0$), one finds again the equation (7.21).

Sinusoidal plane wave rectilineraly polarized

Consider $\underline{G'} = G'_m e^{i'(\omega t - \mathbf{k}\cdot\mathbf{r})}$ with the constant bivector $G'_m = -\mathbf{H}_m + iv\mathbf{D}_m$ and $v = \frac{1}{\sqrt{\varepsilon\mu}}$. In the absence of a four-current density C', one has

$$\begin{aligned}
\nabla'\underline{G'} &= i'\left(j\frac{\omega}{v} + k\mathbf{k}\right)\underline{G'} \\
&= i'\left\{\begin{array}{l} j\left(-v\mathbf{k}\cdot\mathbf{D}\right) + k\left(\mathbf{k}\cdot\mathbf{H}\right) \\ +j\left(v\mathbf{k}\times\mathbf{D} - \frac{\mathbf{H}\omega}{v}\right) + k\left(-\mathbf{k}\times\mathbf{H} - \mathbf{D}\omega\right) \end{array}\right\} \\
&= 0;
\end{aligned}$$

hence, the relations characterizing the plane wave

$$\begin{aligned}
\mathbf{k}\cdot\mathbf{D} &= 0, \\
\mathbf{k}\cdot\mathbf{H} &= 0, \\
\mathbf{H} &= v^2\frac{\mathbf{k}\times\mathbf{D}}{\omega}, \\
\mathbf{D} &= \frac{-\mathbf{k}\times\mathbf{H}}{\omega}.
\end{aligned}$$

7.4 Relativistic optics

7.4.1 Fizeau experiment (1851)

In the Fizeau experiment (Figure 7.2), one observes at O a system of interference fringes when the water is at rest. When the water circulates, the fringes move and the experiment has determined the velocity of light in the moving water to be

$$v = \frac{c}{n} \pm \left(1 - \frac{1}{n^2}\right)w$$

whereas the classical theory predicts

$$v = \frac{c}{n} \pm w.$$

Classical theory

The variation of the optical path δ between the two experiments is

$$\begin{aligned}
\Delta\delta &= c\left[\frac{L}{\left(\frac{c}{n} - w\right)} - \frac{L}{\left(\frac{c}{n} + w\right)}\right] \\
&= \frac{2Lcw}{\frac{c^2}{n^2} - w^2} \simeq \frac{2Ln^2w}{c}.
\end{aligned}$$

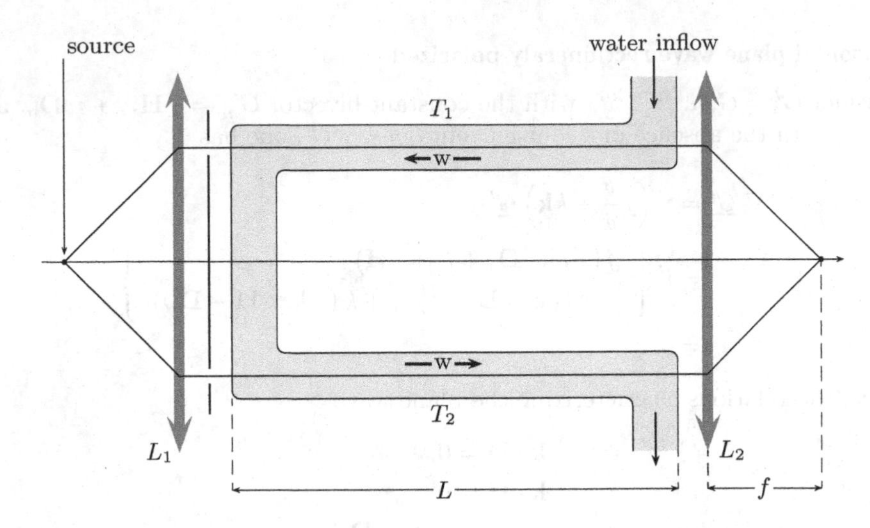

Figure 7.2: Fizeau experiment (1851): when the water is at rest, one observes interference fringes; when the water circulates, the fringes move which allows one to determine the velocity of light in the moving water.

The variation of the order of interference is

$$\Delta p_c = \frac{\Delta \delta}{\lambda_0} = \frac{2Ln^2 w}{c\lambda_0}.$$

Relativistic theory

Let K' be the proper frame of the water in motion with $\frac{c}{n}$ the speed of light with respect to K'. With respect to the frame at rest K, the velocities of light with respect to the tubes T_1, T_2 are respectively using the formula (6.11),

$$v_1 = \frac{\frac{c}{n} - w}{1 - \frac{w}{nc}} \simeq \frac{c}{n} - \left(1 - \frac{1}{n^2}\right) w,$$

$$v_2 = \frac{\frac{c}{n} + w}{1 + \frac{w}{nc}} \simeq \frac{c}{n} + \left(1 - \frac{1}{n^2}\right) w,$$

hence, a variation of the optical path δ,

$$\begin{aligned}\Delta \delta &= c\left(\frac{L}{v_1} - \frac{L}{v_2}\right) \\ &= \frac{2Lcw\left(n^2 - 1\right)}{c^2 - n^2 w^2} \simeq \frac{2Lw}{c}\left(n^2 - 1\right);\end{aligned}$$

the variation of the order of interference is

$$\Delta p_r = \frac{\Delta \delta}{\lambda_0} = \frac{2Lw}{c\lambda_0}\left(n^2 - 1\right).$$

Example. Take $w = 10$ m/s, $L = 5$ m, $\lambda_0 = 0,6$ μm, $n = 4/3$, one obtains

$$\Delta p_c = 0,99; \qquad \Delta p_r = 0,43.$$

7.4.2 Doppler effect

Consider an optical source at rest within the proper frame R' moving with a velocity w with respect to the frame at rest R (Figure 7.3).

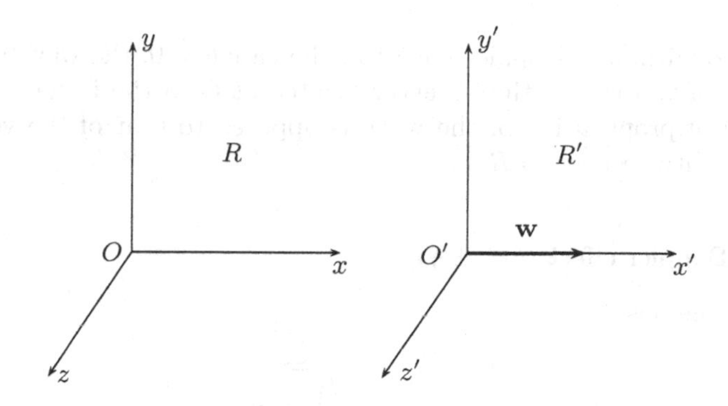

Figure 7.3: Doppler effect: an optical source (at rest in R'), of frequency f', located at O' moves with a velocity w along the Ox axis. In the reference frame R, the measured frequency is f.

Let $K' = j\frac{\omega'}{c} + k\mathbf{k}'$ be the wave four-vector in R' and $K = j\frac{\omega}{c} + k\mathbf{k}$ the wave four-vector in R. One has with $k'_x = k' \cos\theta$ and f, f' designating the frequencies

$$\omega = \gamma\left(\omega' + k'_x \frac{w}{c}\right) = \gamma\omega'\left(1 + \cos\theta\frac{w}{c}\right),$$

$$f = \gamma f'\left(1 + \cos\theta\frac{w}{c}\right).$$

Longitudinal Doppler effect

1. $\theta = 0$ (the source moves towards the observer located at O, Figure 7.4)

$$f = \gamma f'\left(1 + \frac{w}{c}\right) = f'\sqrt{\frac{1 + \frac{w}{c}}{1 - \frac{w}{c}}} \geq f';$$

2. $\theta = \pi$ (the source recedes from the observer, Figure 7.4)

$$f = \gamma f'\left(1 - \frac{w}{c}\right) = f'\sqrt{\frac{1 - \frac{w}{c}}{1 + \frac{w}{c}}} \leq f'.$$

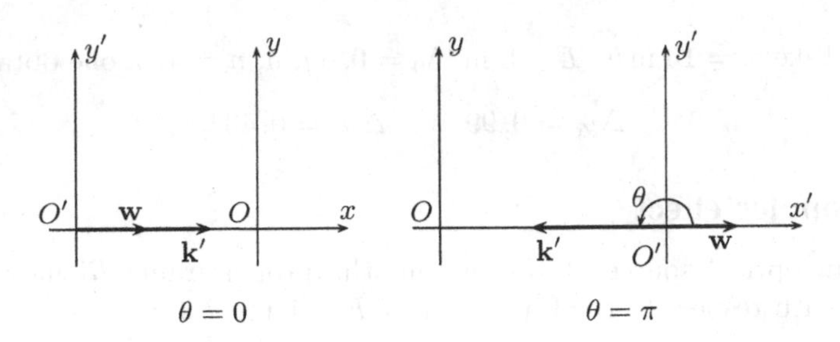

Figure 7.4: Longitudinal Doppler effect: in the case $\theta = 0$, the wave is emitted in the direction of $\mathbf{w}$ towards the observer located at O in the frame R; for $\theta = \pi$, the direction of propagation of the wave is opposed to that of the velocity $\mathbf{w}$ of the frame R' with respect to R.

Transversal Doppler effect ($\theta = \pm\frac{\pi}{2}$)

In this case, one has

$$f = \gamma f' = \frac{f'}{\sqrt{1 - \frac{w^2}{c^2}}}.$$

7.4.3 Aberration of distant stars

With respect to the reference frame R of the sun, let $\mathbf{v}\,(v_x = 0, v_y = -c, v_z = 0)$ be the velocity of light coming from a distant star located on the Oy axis (Figure 7.5).

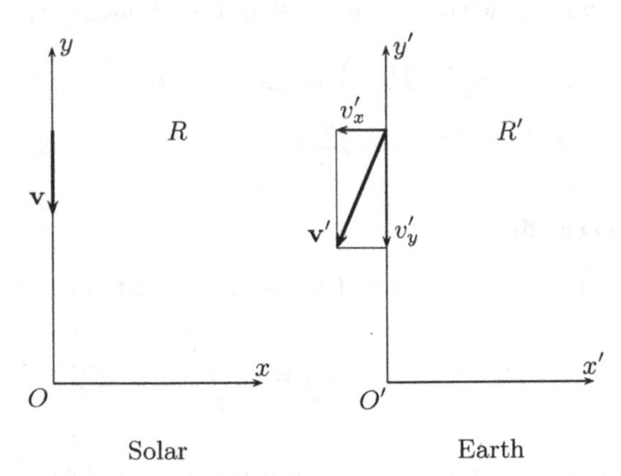

Figure 7.5: Aberration of distant stars: in the reference frame R of the sun, the velocity of light coming from a star located on the Oy axis is $\mathbf{v}$; in the reference frame R' of the Earth having a velocity $\mathbf{w}$ with respect to R, the velocity of light is $\mathbf{v}'$.

The velocity $\mathbf{v}'$ in the mobile reference frame of the Earth is given by equation (6.8)

$$v'_x = \frac{v_x - w}{1 - \frac{v_x w}{c^2}} = -w,$$

$$v'_y = \frac{v_y\sqrt{1 - \frac{w^2}{c^2}}}{1 - \frac{v_x w}{c^2}} = -c\sqrt{1 - \frac{w^2}{c^2}},$$

$$v'_z = \frac{v_z\sqrt{1 - \frac{w^2}{c^2}}}{1 - \frac{v_x w}{c^2}} = 0.$$

The angle θ under which one sees the star from the Earth is given by

$$\tan\theta = \left|\frac{v'_x}{v'_y}\right| = \frac{w}{c\sqrt{1 - \frac{w^2}{c^2}}} = \gamma\frac{w}{c}.$$

7.5 Exercises

E7-1 Consider a charge q located at the origin of a reference frame K' moving with a constant velocity v along the Ox axis of a reference frame K at rest. Determine the four-potential and the electromagnetic field in the frame K at rest.

E7-2 Consider an infinite linear distribution of charges with a linear density λ_0 ($\lambda_0 > 0$) along the $O'x'$ axis of a reference frame K' moving with a velocity v along the Ox axis of a reference frame K at rest. Along the Ox axis of the reference frame K, one has an infinite linear distribution with a linear density $-\lambda$ ($\lambda > 0$) of charges at rest. Knowing that the total linear density of charges in the reference frame K at rest is nil, determine the electromagnetic field created in K by these charges at the point $M'(x', r', 0)$. A particle of charge q moves with a velocity v in K, parallel to the Ox axis (at the distance r). Determine the force exerted by the two sets of charges in the reference frame K and in the proper frame of the particle.

E7-3 Consider the electromagnetic field

$$F = -I + 2\frac{iJ}{c} \qquad (B_x = 1T,\ E_y = 2V/m)$$

in the reference frame of the laboratory K at rest . Determine the electromagnetic field in the reference frame moving with respect to K with a velocity $\frac{c}{2}$ in the direction $(1,1,0)$ of K.

E7-4 Show that Maxwell's equations in vacuum without sources are invariant under the transformation

$$F \to F^* = iF \qquad \left(\text{i.e., } \mathbf{B} \to \frac{\mathbf{E}}{c},\ \frac{\mathbf{E}}{c} \to -\mathbf{B}\right).$$

Chapter 8

General relativity

The general theory of relativity is developed within the Clifford algebra $\mathbb{H}\otimes\mathbb{H}$ over $\mathbb{R}$. Einstein's equations and the equation of motion are given as well as applications such as the Schwarzschild metric and the linear approximation.

8.1 Riemannian space

Consider a four-dimensional space with the elementary displacement $DM = \omega^i e_i$ and the affine connection $De_i = \omega_i^j e_j$. The covariant differentiation of the vector A is defined by

$$DA = dA + d\omega \cdot A$$

with $2d\omega = \omega^{ij} e_i \wedge e_j$ et $\omega^{ij} = \omega_k^i(e^k \cdot e^j)$. The reciprocal basis e^α is defined by $e^\mu \cdot e_v = \delta_v^\mu$ ($e^0 = e_0$, $e^1 = -e_1$, $e^2 = -e_2$, $e^3 = -e_3$) where e_0, e_1, e_2, e_3 are unitary orthogonal vectors. Under a Lorentz transformation $A' = fAf_c$, one has

$$DA' = fDAf_c,$$
$$d\omega' = fd\omega f_c - 2df\, f_c.$$

A Riemannian space is a space without torsion but with a curvature. The absence of torsion is expressed by

$$D_2(D_1 M) - D_1(D_2 M) = 0 \tag{8.1}$$

where D_1, D_2 are two linearly independent directions. Writing $d\omega = I_k dx^k, DM = \sigma_m dx^m$, condition (8.1) leads to the relations

$$I_k \cdot \sigma_m - I_m \cdot \sigma_k = \frac{\partial \sigma_k}{\partial x^m} - \frac{\partial \sigma_m}{\partial x^k}$$

which determine I_k when DM is given. The existence of a curvature is expressed by the relations

$$(D_2 D_1 - D_1 D_2)A = \Omega \cdot A$$

where Ω is a bivector defined by

$$\begin{aligned}\Omega &= (d_2 d_1 - d_1 d_2)\omega + [d_2\omega, d_1\omega] \\ &= \Omega_{km} dS^{km}/2 \\ &= \left\{ \frac{\partial I_k}{\partial x^m} - \frac{\partial I_m}{\partial x^k} - [I_k, I_m] \right\} d_1 x^k d_2 x^m\end{aligned}$$

with $2\Omega_{km} = -R^{ij}_{km} e_i \wedge e_j$ and $dS = D_1 M \wedge D_2 M = dS^{km} e_k \wedge e_m/2$. Bianchi's first identity is given by

$$\Omega_{ij} \cdot e_k + \Omega_{jk} \cdot e_i + \Omega_{ki} \cdot e_j = 0, \tag{8.2}$$

$$\Omega_{ij} \cdot (e_k \wedge e_m) = \Omega_{km} \cdot (e_i \wedge e_j), \tag{8.3}$$

where relation (8.3) results from the preceeding equation. Bianchi's second identity is expressed by

$$\Omega_{ij;k} + \Omega_{jk;i} + \Omega_{ki;j} = 0$$

with $D_3 \Omega(d_2, d_1) = d_3 \Omega + [d_3 \omega, \Omega] = \Omega_{ij;k} \omega^i (d_2) \omega^j (d_1) \omega^k (d_3)$. The Ricci tensor $R_{ik} = R^h_{ihk}$ and the curvature $R = R^k_k$ are obtained by the relations

$$\begin{aligned}R_k &= \Omega_{ik} \cdot e^i = R_{ik} e^i, \\ R &= (\Omega_{ik} \cdot e^i) \cdot e^k = \Omega_{ik} \cdot (e^i \wedge e^k).\end{aligned}$$

8.2 Einstein's equations

To deduce Einstein's equations from a variational principle, we shall use the method of the exterior calculus but within the framework of a Clifford algebra and without using the exterior product of differential forms. Adopting, for simplicity, an orthogonal curvilinear coordinate system, we can write

$$\begin{aligned}L &= \int R\sqrt{-g}\, dx^0 dx^1 dx^2 dx^3 \\ &= \int \Omega_{ik} \cdot (e^i \wedge e^k) \omega^0 \omega^1 \omega^2 \omega^3\end{aligned}$$

with $\sqrt{-g}\, dx^0 dx^1 dx^2 dx^3 = \omega^0 \omega^1 \omega^2 \omega^3$. Taking the dual, we obtain

$$L^* = \int \Omega_{ik} \wedge (e^i \wedge e^k)^* \omega^0 \omega^1 \omega^2 \omega^3.$$

The variation gives

$$\begin{aligned}\delta L^* &= \int \Omega_{ik} \omega^I \omega^K \wedge \delta \left[(e^i \wedge e^k)^* \omega^G \omega^H \right] \\ &\quad + \int \delta \left(\Omega_{ik} \omega^I \omega^K \right) \wedge (e^i \wedge e^k)^* \omega^G \omega^H\end{aligned}$$

with $i = I$, $k = K$ and where the second integral vanishes. Furthermore, one has

$$\begin{aligned}\delta\left[\left(e^i \wedge e^k\right)^*\right] &= \delta\left(e^\gamma\right) \wedge \left[e_\gamma \cdot \left(e^i \wedge e^k\right)^*\right] \\ &= -\delta\left(e^\gamma\right) \wedge \left[\left(e^i \wedge e^k \wedge e_\gamma\right)^*\right] \\ &= -\delta\left(e_\gamma\right) \wedge \left[\left(e^i \wedge e^k \wedge e^\gamma\right)^*\right].\end{aligned}$$

Hence

$$\begin{aligned}\delta L^* &= \int \Omega_{ik}\omega^I\omega^K \wedge \left[\delta\left(\omega^g e_\gamma\right) \wedge \left(e^i \wedge e^k \wedge e^\gamma\right)^* \omega^h\right] \\ &= -\int \delta\left(\omega^g e_\gamma\right) \wedge \left[\Omega_{ik} \wedge \left(e^i \wedge e^k \wedge e^\gamma\right)^* \omega^I\omega^K\omega^h\right] = 0\end{aligned}$$

with $g = \gamma$, $i = I$, $k = K$ (without summation). Finally, one obtains Einstein's equations in a vacuum

$$\Omega_{ij} \wedge \left(e^i \wedge e^j \wedge e^k\right)^* = 0.$$

In the presence of an energy-momentum distribution, one has

$$\frac{1}{2}\Omega_{ij} \wedge \left(e^i \wedge e^j \wedge e^k\right)^* = \varkappa\left(T^k\right)^* \tag{8.4}$$

with $T^k = T^{ik}e_i, \varkappa = 8\pi G/c^4$ where T^{ik} is the energy-momentum tensor ([50], [23], [34]). The standard expression of Einstein's equations is obtained as follows. Taking the dual of equation (8.4), one obtains

$$-\frac{1}{2}\Omega_{ij} \cdot \left(e^i \wedge e^j \wedge e^k\right) = \varkappa\left(T^k\right).$$

Using the formula $(B \cdot T) \cdot V = (B \wedge V) \cdot T$ where B, T, V are respectively a bivector, a trivector and a vector, one has

$$\begin{aligned}-\left[\frac{1}{2}\Omega_{ij} \cdot \left(e^i \wedge e^j \wedge e^k\right)\right] \cdot e_\mu &= \frac{1}{2}R_{ij}^{\alpha\beta}\left(e_\alpha \wedge e_\beta \wedge e_\mu\right) \cdot \left(e^i \wedge e^j \wedge e^k\right) \\ &= -\frac{1}{2}R_{ij}^{\alpha\beta}\delta_{\alpha\beta\mu}^{ijk} = R_{\beta\mu}^{\beta k} - \frac{1}{2}\delta_\mu^k R \\ &= \varkappa T^k \cdot e_\mu = \varkappa T^{ik}e_i \cdot e_\mu = \varkappa T_\mu^k.\end{aligned}$$

8.3 Equation of motion

To obtain the equation of motion, one writes

$$\delta S = -mc\int \delta\left(ds\right) = 0$$

with $\delta(ds) = \delta(ds)^2/2ds$ and $ds^2 = DM \cdot DM = (\sigma_i \cdot \sigma_k)\, dx^i dx^k$. One obtains

$$\begin{aligned}\delta S &= -mc \int (\delta\sigma_i \cdot \sigma_k)\, u^i u^k ds - mc \int \left[(\sigma_l \cdot \sigma_k)\, u^k\right] d\delta x^l \\ &= A + B = 0\end{aligned}$$

with $u^i = dx^i/ds$. The second term B can be rewritten in the form

$$B = -mc \int \left[(\sigma_l \cdot \sigma_k)\, u^k\right] \delta x^l + mc \int \delta x^l d\left[(\sigma_l \cdot \sigma_k)\, u^k\right].$$

Since, $\delta\sigma_i = \sigma_{i,l}\delta x^l + \delta\omega \cdot \sigma_i$ with $\delta\omega = I_k dx^k$, one obtains

$$-\left[\frac{\partial\sigma_i}{\partial x^l} + I_l \cdot \sigma_i\right] \cdot \sigma_k u^i u^k + \left[\frac{d\,(\sigma_l \cdot \sigma_k)}{ds}\right] u^k + (\sigma_l \cdot \sigma_k)\, \frac{du^k}{ds} = 0$$

or, equivalently,

$$\left\{\begin{array}{c}\left[-\dfrac{\partial\sigma_i}{\partial x^l} + \dfrac{\partial\sigma_l}{\partial x^i} - I_l \cdot \sigma_i + I_i \cdot \sigma_l\right] \cdot \sigma_k u^k u^i \\ +\sigma_l \cdot \left(\dfrac{\partial\sigma_k}{\partial x^i} + I_i \cdot \sigma_k\right) u^i u^k + (\sigma_l \cdot \sigma_k)\, \dfrac{du^k}{ds}\end{array}\right\} = 0$$

where the expression in brackets vanishes in the absence of torsion. One thus obtains

$$\sigma_l \cdot \left(\frac{du}{ds} + I_i u^i \cdot u\right) = 0$$

with the four-velocity $u = DM/ds = \sigma_i u^i$ and the ordinary derivative du/ds. Finally, the equation of motion is expressed by

$$\frac{Du}{ds} = \frac{du}{ds} + \frac{d\omega}{ds} \cdot u = 0.$$

8.4 Applications

8.4.1 Schwarzschild metric

Consider a metric having a spherical symmetry and the elementary displacement DM,

$$\begin{aligned}ds^2 &= e^{2a}c^2dt^2 - e^{2b}dr^2 - r^2d\theta^2 - r^2(\sin\theta)^2d\varphi^2, \\ DM &= e^a cdte_0 + e^b dre_1 + rd\theta e_2 + r\sin\theta d\varphi e_3\end{aligned}$$

where a, b are functions of t and r. The bivector $d\omega$ is given by

$$d\omega = \cos\theta d\varphi I - e^{-b}\sin\theta d\varphi J + e^{-b}d\theta K + \left(a'e^{a-b} + \frac{\dot{b}}{c}e^{-a+b}\right) iI$$

where the prime and the point designate respectively a partial derivative with respect to r and t. The bivectors Ω_{ik} are given by

$$\Omega_{01} = \left\{\left[a'' + (a')^2 - a'b'\right] e^{-2b} + \left[-\ddot{b} - \left(\dot{b}\right)^2 + \dot{b}\dot{a}\right] \frac{e^{-2a}}{c^2}\right\} iI \quad ,$$

$$\Omega_{02} = \frac{\dot{b}e^{-a-b}}{rc} K + \frac{a'e^{-2b}}{r} iJ, \qquad \Omega_{03} = -\frac{\dot{b}e^{-a-b}}{rc} J + \frac{a'e^{-2b}}{r} iK,$$

$$\Omega_{12} = \frac{b'e^{-2b}}{r} K + \frac{\dot{b}}{rc} e^{-a-b} iJ, \qquad \Omega_{13} = -\frac{b'e^{-2b}}{r} J + \frac{\dot{b}}{rc} e^{-a-b} iK,$$

$$\Omega_{23} = \frac{\left(1 - e^{-2b}\right)}{r^2} I.$$

Einstein's equations, in a vacuum, are

$$\begin{aligned} \Omega_{23} \wedge e_1 - \Omega_{13} \wedge e_2 + \Omega_{12} \wedge e_3 &= 0, \\ -\Omega_{23} \wedge e_0 + \Omega_{03} \wedge e_2 - \Omega_{02} \wedge e_3 &= 0, \\ \Omega_{13} \wedge e_0 - \Omega_{03} \wedge e_1 + \Omega_{01} \wedge e_3 &= 0, \\ -\Omega_{12} \wedge e_0 + \Omega_{02} \wedge e_1 - \Omega_{01} \wedge e_2 &= 0. \end{aligned}$$

Developing these four equations, one obtains respectively

$$-\left(\frac{1 - e^{-2b}}{r^2} + \frac{2b'e^{-2b}}{r}\right) k - \frac{2\dot{b}}{cr} e^{-a-b} jI = 0, \tag{8.5}$$

$$-\left(\frac{1 - e^{-2b}}{r^2} - \frac{2a'e^{-2b}}{r}\right) jI + \frac{2\dot{b}}{cr} e^{-a-b} k = 0, \tag{8.6}$$

$$-\left\{\begin{aligned} &\frac{b' - a'}{r} e^{-2b} \\ &+\left[(\ddot{b} + \dot{b}^2 - \dot{b}\dot{a})\frac{e^{-2a}}{c^2} - (a'' + a'^2 - a'b')e^{-2b}\right] \end{aligned}\right\} jJ = 0, \tag{8.7}$$

$$-\left\{\begin{aligned} &\frac{b' - a'}{r} e^{-2b} \\ &+\left[(\ddot{b} + \dot{b}^2 - \dot{b}\dot{a})\frac{e^{-2a}}{c^2} - (a'' + a'^2 - a'b')e^{-2b}\right] \end{aligned}\right\} jK = 0. \tag{8.8}$$

From equations (8.5), (8.6), one obtains $\dot{b} = 0$ and $a' + b' = 0$ or equivalently $\log e^{2(a+b)} = C$ where $C = 0$ because $e^{2a} = e^{2b} = 1$ at infinity; hence $a + b = 0$. Equation (8.5) then gives

$$e^{-2b} \left(2b'r - 1\right) + 1 = 0,$$

$$\left(re^{-2b}\right)' = 0,$$

$$e^{-2b} = 1 + \frac{C_1}{r},$$

and taking $C_1 = -2m$ (with $m = GM/c^2$), one has $e^{2b} = e^{-2a} = \left(1 - \frac{2m}{r}\right)^{-1}$ which determines completely the Schwarzschild metric. Finally, one has

$$ds^2 = \left(1 - \frac{2m}{r}\right) c^2 dt^2 - \left(1 - \frac{2m}{r}\right)^{-1} dr^2 - r^2 d\theta^2 - r^2 \sin^2\theta d\varphi^2,$$

$$d\omega = \cos\theta d\varphi I + \left(1 - \frac{2m}{r}\right)^{1/2} (-\sin\theta d\varphi J + d\theta K) + \frac{mcdt}{r^2} iI,$$

$$\Omega_{01} = -\frac{2m}{r^3} iI, \qquad \Omega_{02} = \frac{m}{r^3} iJ, \qquad \Omega_{03} = \frac{m}{r^3} iK,$$

$$\Omega_{12} = -\frac{m}{r^3} K, \qquad \Omega_{13} = \frac{m}{r^3} J, \qquad \Omega_{23} = \frac{2m}{r^3} I.$$

The equation of motion is expressed simply in the form

$$\frac{Du}{ds} = \frac{du}{ds} + \frac{d\omega}{ds} \cdot u = 0.$$

Developing this expression along the axes e_0, e_1, e_2, e_3 one obtains respectively

$$\frac{d}{ds}\left[\left(1 - \frac{2m}{r}\right)\frac{cdt}{ds}\right] = 0, \tag{8.9}$$

$$\frac{d}{ds}\left[\left(1 - \frac{2m}{r}\right)^{-\frac{1}{2}} \frac{dr}{ds}\right] = \left(1 - \frac{2m}{r}\right)^{\frac{1}{2}} \left[\begin{array}{c} -\frac{m}{r^2}\left(\frac{cdt}{ds}\right)^2 + r\left(\frac{d\theta}{ds}\right)^2 \\ +r\sin^2\theta\left(\frac{d\varphi}{ds}\right)^2 \end{array}\right], \tag{8.10}$$

$$\frac{d}{ds}\left(r^2 \frac{d\theta}{ds}\right) = r^2 \sin\theta \left(\frac{d\varphi}{ds}\right)^2, \tag{8.11}$$

$$\frac{d}{ds}\left(r^2 \sin^2\theta \frac{d\varphi}{ds}\right) = 0. \tag{8.12}$$

Adopting $\theta = \pi/2$, it follows from the equations (8.9), (8.12) that

$$\left(1 - \frac{2m}{r}\right)\frac{cdt}{ds} = k, \qquad r^2 \frac{d\varphi}{ds} = h,$$

where k and h are constants. The relation $u_c u = 1$ is expressed by

$$\left(1 - \frac{2m}{r}\right)\left(\frac{cdt}{ds}\right)^2 - \left(1 - \frac{2m}{r}\right)^{-1}\left(\frac{dr}{ds}\right)^2 - r^2\left(\frac{d\varphi}{ds}\right)^2 = 1,$$

or with $w = 1/r$,

$$\frac{dr}{ds} = -h\frac{dw}{d\varphi}, \qquad \frac{d^2r}{ds^2} = -h^2 w^2 \frac{d^2w}{d\varphi^2}.$$

The projection of the equation on the axis e_1 then leads to the relation

$$\frac{d^2w}{d\varphi^2} + w = \frac{m}{h^2} + 3mw^2$$

which is the relativistic form of Binet's equation and from which one can deduce the precession of the perihelion of Mercury in the usual way ([34]), ([46]).

8.4.2 Linear approximation

Consider the elementary displacement $DM = \omega^i e_i$ with $\omega^i = dx^i + h^i_j dx^j$ where $h_{ij} = h_{ji} = \eta_{ik} h^k_j$ is small as compared to 1 [50, p. 186]. The bivectors $d\omega$ and Ω_{km} are given by

$$2d\omega = (h_{ik,j} - h_{jk,i}) e^i \wedge e^j dx^k,$$
$$2\Omega_{km} = (h_{ik,jm} - h_{im,jk}) e^i \wedge e^j.$$

Using the relation $(e_i \wedge e_j) \cdot e^J = e_i$ (if $J = j \neq i$) and nil otherwise, one obtains, in a vacuum, with harmonic coordinates ($2h_{mi,i} = h_{ii,m}$),

$$R_k = \Omega_{km} \cdot e^i = -(\Box h^m_k) e_m = 0,$$

the last equation being a gravitational wave equation (with $\Box = \frac{\partial^2}{c^2 \partial t^2} - \frac{\partial^2}{\partial x^2} - \frac{\partial^2}{\partial y^2} - \frac{\partial^2}{\partial z^2}$). Writing

$$\gamma_{km} = h_{km} - \frac{1}{2} \eta_{km} h$$

with $h = h^m_m$ one obtains Einstein's equations

$$-\Box \gamma^k = \varkappa T^k$$

(with $\gamma^k = \gamma^{mk} e_m$).

If one considers a homogeneous sphere (of mass M and radius r_0) rotating slowly with an angular velocity ω around the axis Oz , one integrates the above equations and obtains the metric (with $m = GM/c^2$) for $r \geq r_0$,

$$ds^2 = \left(1 - \frac{2m}{r}\right) c^2 dt^2 - \left(1 + \frac{2m}{r}\right) \left(dx^2 + dy^2 + dz^2\right) - \frac{4GI\omega}{c^3 r^3} (ydx - xdy)\, cdt$$

with $I = 2Mr_0^2/5$. The elementary displacement is

$$DM = \omega^i e_i = (dx^i + h^i_j dx^j) e_i$$

with

$$h_{00} = h_{11} = h_{22} = h_{33} = -m/r,$$
$$h_{01} = -GI\omega y / c^3 r^3, \qquad h_{02} = -GI\omega x / c^3 r^3,$$

the other h_{ij} being equal to zero. The bivector $d\omega = \frac{1}{2}(h_{ik,j} - h_{jk,i}) e^i \wedge e^j dx^k$ gives the equation of motion

$$\frac{Du}{ds} = \left[\begin{array}{c} -j\,(u_1 h_{00,1} + u_2 h_{00,2} + u_3 h_{00,3}) \\ +kI\,(-h_{00,1} + u_3 \Omega_2 - u_2 \Omega_3) \\ +kJ\,(-h_{00,2} + u_1 \Omega_3 - u_3 \Omega_1) \\ +kK\,(-h_{00,3} + u_2 \Omega_1 - u_1 \Omega_2) \end{array} \right]$$

with $h_{00,1} = mx/r^3$, $h_{00,2} = my/r^3$, $h_{00,3} = mz/r^3$ and

$$\Omega_1 = h_{30,2} - h_{20,3} = 3GI\omega xz/c^3r^5,$$
$$\Omega_2 = h_{10,3} - h_{30,1} = 3GI\omega yz/c^3r^5,$$
$$\Omega_3 = h_{20,1} - h_{10,2} = GI\omega\left(2z^2 - x^2 - y^2\right)/c^3r^5,$$

which puts in evidence the Thirring precession with respect to the local inertial frame.

Conclusion

From the abstract quaternion group, we have defined the quaternion algebra $\mathbb{H}$, then the complex quaternion algebra $\mathbb{H}(\mathbb{C})$ and the Clifford algebra $\mathbb{H} \otimes \mathbb{H}$. The quaternion algebra gives a representation of the rotation group SO(3) well-known for its simplicity and its immediate physical significance. The Clifford algebra $\mathbb{H}\otimes\mathbb{H}$ yields a double representation of the Lorentz group containing the SO(3) group as a particular case and having also an immediate physical meaning. Furthermore, the algebra $\mathbb{H} \otimes \mathbb{H}$ constitutes the framework of a relativistic multivector calculus, equipped with an associative exterior product and interior products generalizing the classical vector and scalar products. This calculus remains relatively close to the classical vector calculus which it contains as a particular case. The Clifford algebra $\mathbb{H} \otimes \mathbb{H}$ allows us to easily formulate special relativity, classical electromagnetism and general relativity. In complexifying $\mathbb{H} \otimes \mathbb{H}$, one obtains the Dirac algebra, Dirac's equation, relativistic quantum mechanics and a simple formulation of the unitary group SU(4) and the symplectic unitary group $\mathrm{USp}(2, \mathbb{H})$. Algebraic or numerical computations within the Clifford algebra $\mathbb{H} \otimes \mathbb{H}$ have become straightforward with software such as *Mathematica*. We hope to have shown that the Clifford algebras $\mathbb{H} \otimes \mathbb{H}$ over $\mathbb{R}$ and $\mathbb{C}$ constitute a coherent, unified, framework of mathematical tools for special relativity, classical electromagnetism, general relativity and relativistic quantum mechanics. The quaternion group consequently appears via the Clifford algebra as a fundamental structure of physics revealing its deep harmony.

Appendix A

Solutions

Chapter 1

S1-1

$$i^2 = j^2 = k^2 = ijk = -1,$$

$$(-i)\,ijk = jk = i,$$
$$k = -ji,$$

$$ijk\,(-k) = ij = k,$$
$$j = -ik,$$
$$i = -kj, \text{ etc.}$$

S1-2

$$|a| = \sqrt{2}, \qquad |b| = \sqrt{25} = 5, \qquad a^{-1} = \frac{1}{2}(1-i), \qquad b^{-1} = \frac{1}{5}(4+3j),$$
$$a + b = 5 + i - 3j, \qquad ab = 4 + 4i - 3j - 3k, \qquad ba = 4 + 4i - 3j + 3k,$$
$$a = \sqrt{2}\frac{1+i}{\sqrt{2}} = \sqrt{2}\left(\cos 45^\circ + i \sin 45^\circ\right),$$
$$b = 5\left(\frac{4-3j}{5}\right) = 5\left(\cos 36,87^\circ - \sin 36,87^\circ\right),$$
$$\sqrt{a} = 2^{1/2}\left(\cos\frac{45}{2} + i\sin\frac{45}{2}\right), \qquad \sqrt{b} = 5^{1/2}\left(\cos\frac{36,87}{2} - i\sin\frac{36,87}{2}\right),$$
$$a^{1/3} = 2^{1/6}\left(\cos\frac{45}{3} + i\sin\frac{45}{3}\right) = 2^{1/6}\left(\cos 15^\circ + i\sin 15^\circ\right).$$

S1-3

$$a_c a x + a_c x b = a_c c,$$
$$a x b + x b^2 = c b,$$
$$x\left[a a_c + b^2 + 2S(a)b\right] = a_c c + c b_c;$$

the latter is a equation of the type $x\alpha = \beta, \alpha, \beta \in H$ which one solves in x.
Answer: $2ix + xj = k, \qquad x = \frac{1}{3}(2j - i).$

S1-4

$$\begin{aligned} axbb^{-1} + cxdb^{-1} &= eb^{-1}, \\ c^{-1}ax + xdb^{-1} &= c^{-1}eb^{-1}; \end{aligned}$$

one finds an equation of the type of the previous exercice and solves similarly;
Answer: $x = 2 + i$.

S1-5

$$\begin{aligned} x^{-1}x^2 &= x^{-1}xa + x^{-1}bx, \\ x &= a + x^{-1}bx, \\ xx^{-1} &= ax^{-1} + x^{-1}b = 1; \end{aligned}$$

Answer: $x = \frac{1}{5}(-6j + 3k.)$.

Chapter 2

S2-1

$$A_\alpha = \begin{bmatrix} \cos\alpha & -\sin\alpha & 0 \\ \sin\alpha & \cos\alpha & 0 \\ 0 & 0 & 1 \end{bmatrix}, \qquad A_\beta = \begin{bmatrix} \cos\beta & 0 & \sin\beta \\ 0 & 1 & 0 \\ -\sin\beta & 0 & \cos\beta \end{bmatrix},$$

$$A_\gamma = \begin{bmatrix} 1 & 0 & 0 \\ 0 & \cos\gamma & -\sin\gamma \\ 0 & \sin\gamma & \cos\gamma \end{bmatrix}.$$

S2-2

$$\begin{aligned} dA &= rdA'r_c + drA'r_c + rA'dr_c \\ &= r\left(dA' + r_c drA' + A'dr_c r\right) r_c\,; \end{aligned}$$

since $r_c dr = -dr_c r$ (which is obtained by differentiating $rr_c = 1$), one obtains with $d\Omega' = 2r_c dr$,

$$\begin{aligned} dA &= r\left(dA' + r_c drA' - A'r_c dr\right) r_c \\ &= r\left(dA' + \frac{d\Omega'}{2}A' - A'\frac{d\Omega'}{2}\right) r_c = r\left(dA' + d\Omega' \times A'\right) r_c = rDA'r_c; \\ DA' &= dA' + d\Omega' \times A'\,; \end{aligned}$$

$\frac{d\Omega'}{dt}$ is the angular velocity with respect to the moving frame.

S2-3

$$\begin{aligned}
A'' &= fA'f_c \;(f = g_c r, ff_c = 1),\\
DA'' &= fDA'f_c,\\
dA'' + d\Omega'' \times A'' &= f\,(dA' + d\Omega' \times A')\,f_c,\\
(df A' f_c + f dA' f_c + f A' df_c) + d\Omega'' \times A'' &= f dA' f_c + f\,(d\Omega' \times A')\,f_c,
\end{aligned}$$

$$\begin{aligned}
d\Omega'' \times A'' &= f\,(d\Omega' \times A')\,f_c - df\,f_c\,(fA'f_c) - (fA'f_c)\,f df_c\\
&= f d\Omega' f_c \times A'' - df\,f_c A'' + A'' df\,f_c\\
&= (f d\Omega' f_c - 2df\,f_c) \times A'';
\end{aligned}$$

hence

$$\begin{aligned}
d\Omega'' &= f d\Omega' f_c - 2df\,f_c,\\
d\Omega' &= f_c d\Omega'' f - 2f_c df.
\end{aligned}$$

S2-4

$$\begin{aligned}
&X = rX'r_c,\\
&r = \cos\frac{\theta}{2} + k\sin\frac{\theta}{2},\ \frac{d\Omega'}{dt} = 2r_c\frac{dr}{dt} = k\frac{d\theta}{dt}.
\end{aligned}$$

Polar coordinates:

$$\begin{aligned}
X &= xi + yj, \qquad X' = \rho i,\\
V' &= \frac{DX'}{dt} = \frac{dX'}{dt} + \frac{d\Omega'}{dt} \times X' = \frac{d\rho}{dt}i + \rho\frac{d\theta}{dt}j;\\
\gamma' &= \frac{DV'}{dt} = \frac{dV'}{dt} + \frac{d\Omega'}{dt} \times V'\\
&= \left[\frac{d^2\rho}{dt^2} - \rho\left(\frac{d\theta}{dt}\right)^2\right] i + j\left[2\rho\frac{d\rho}{dt}\frac{d\theta}{dt} + \rho\frac{d^2\theta}{dt^2}\right].
\end{aligned}$$

Cylindrical coordinates:

$$\begin{aligned}
X &= xi + yj + zk, \qquad X' = \rho i + zk,\\
V' &= \frac{DX'}{dt} = \frac{dX'}{dt} + \frac{d\Omega'}{dt} \times X' = \frac{d\rho}{dt}i + \rho\frac{d\theta}{dt}j + k\frac{dz}{dt};\\
\gamma' &= \frac{DV'}{dt} = \frac{dV'}{dt} + \frac{d\Omega'}{dt} \times V'\\
&= \left[\frac{d^2\rho}{dt^2} - \rho\left(\frac{d\theta}{dt}\right)^2\right] i + \left[\frac{1}{\rho}\frac{d}{dt}\left(\rho^2\frac{d\theta}{dt}\right)\right] j + \left[\frac{d^2z}{dt^2}\right] k.
\end{aligned}$$

S2-5 The moving frame (ρ, θ, φ) is obtained with the following three rotations:

$$r_1 = \cos\frac{\pi}{4} - j\sin\frac{\pi}{4} \text{ (rotation of } -\tfrac{\pi}{2} \text{ around } Oy),$$
$$r_2 = \cos\frac{\theta}{2} - i\sin\frac{\theta}{2} \text{ (rotation of } -\theta \text{ around } Ox),$$
$$r_3 = \cos\frac{2\varphi - \pi}{4} + k\sin\frac{2\varphi - \pi}{4} \text{ (rotation of } -(\tfrac{\pi}{2} - \varphi) \text{ around } Oz);$$

$$\begin{aligned} r &= r_3 r_2 r_1 \\ &= \frac{1}{2}\left[\cos\frac{\theta+\varphi}{2} + \sin\frac{\theta+\varphi}{2}\right] \\ &\quad + \frac{1}{\sqrt{2}}\left[i\sin\frac{2\varphi - 2\theta - \pi}{4} - j\cos\frac{2\theta - 2\varphi + \pi}{4} + k\sin\frac{2\varphi + 2\theta - \pi}{4}\right]; \end{aligned}$$

$$\begin{aligned} d\Omega' &= 2r_c dr = i d\varphi\cos\theta - j d\varphi\sin\theta + k d\theta, \\ d\Omega &= 2dr r_c = -i d\theta\sin\varphi + j d\theta\cos\varphi + k d\varphi; \end{aligned}$$

$$\begin{aligned} e_1 &= r i r_c = i\sin\theta\cos\varphi + j\sin\theta\sin\varphi + k\cos\theta, \\ e_2 &= r j r_c = i\cos\theta\cos\varphi + j\cos\theta\sin\varphi - k\sin\theta, \\ e_3 &= r k r_c = -i\sin\varphi + j\cos\varphi; \end{aligned}$$

$$\begin{aligned} V' &= \frac{DX'}{dt} = \frac{dX'}{dt} + \frac{d\Omega'}{dt}\times X' = i\frac{d\rho}{dt}i + j\rho\frac{d\theta}{dt} + k\rho\frac{d\varphi}{dt}\sin\theta, \\ \gamma' &= \frac{DV'}{dt} = \frac{dV'}{dt} + \frac{d\Omega'}{dt}\times V' \\ &= i\left\{\frac{d^2\rho}{dt^2} - \rho\left[\left(\frac{d\theta}{dt}\right)^2 + \left(\frac{d\varphi}{dt}\right)^2\sin^2\theta\right]\right\} \\ &\quad + j\left\{\frac{1}{\rho}\frac{d}{dt}\left(\rho^2\frac{d\theta}{dt}\right) - \rho\sin\theta\cos\theta\left(\frac{d\varphi}{dt}\right)^2\right\} \\ &\quad + k\left\{\frac{1}{\rho\sin\theta}\frac{d}{dt}\left(\rho^2\sin^2\theta\frac{d\varphi}{dt}\right)\right\}. \end{aligned}$$

S2-6

$$r_1 = \cos\frac{\alpha}{2} + k\sin\frac{\alpha}{2} \text{ (1}^{\text{st}} \text{ rotation)},$$
$$r_2 = \cos\frac{\beta}{2} + \sin\frac{\beta}{2} r_1 i r_{1c} = r_1\left(\cos\frac{\beta}{2} + i\sin\frac{\beta}{2}\right) r_{1c} \text{ (2}^{\text{nd}} \text{ rotation)},$$
$$r_3 = \cos\frac{\gamma}{2} + \sin\frac{\gamma}{2}(r_2 k r_{2c}) \text{ (3}^{\text{rd}} \text{ rotation)};$$

$$\begin{aligned}
r &= r_3 r_2 r_1 \\
&= r_2 \left(\cos\frac{\gamma}{2} + k\sin\frac{\gamma}{2}\right) r_{2c} r_2 r_1 \\
&= r_1 \left(\cos\frac{\beta}{2} + i\sin\frac{\beta}{2}\right) r_{1c} \left(\cos\frac{\gamma}{2} + k\sin\frac{\gamma}{2}\right) r_1 \\
&= \left(\cos\frac{\alpha}{2} + k\sin\frac{\alpha}{2}\right)\left(\cos\frac{\beta}{2} + i\sin\frac{\beta}{2}\right)\left(\cos\frac{\gamma}{2} + k\sin\frac{\gamma}{2}\right)
\end{aligned}$$

because

$$r_{1c} \left(\cos\frac{\gamma}{2} + k\sin\frac{\gamma}{2}\right) r_1 = \left(\cos\frac{\gamma}{2} + k\sin\frac{\gamma}{2}\right);$$

finally

$$r = \cos\frac{\beta}{2}\cos\frac{\alpha+\gamma}{2} + i\sin\frac{\beta}{2}\cos\frac{\alpha-\gamma}{2} + j\sin\frac{\beta}{2}\sin\frac{\alpha-\gamma}{2} + \cos\frac{\beta}{2}\sin\frac{\alpha+\gamma}{2},$$

$$\begin{aligned}
\omega' = 2r_c\frac{dr}{dt} = {} & i\left(\frac{d\beta}{dt}\cos\gamma + \frac{d\alpha}{dt}\sin\beta\sin\gamma\right) \\
& + j\left(-\frac{d\beta}{dt}\sin\gamma + \frac{d\alpha}{dt}\sin\beta\cos\gamma\right) \\
& + k\left(\frac{d\gamma}{dt} + \frac{d\alpha}{dt}\cos\beta\right);
\end{aligned}$$

$$\begin{aligned}
\omega = 2\frac{dr}{dt}r_c = {} & i\left(\frac{d\beta}{dt}\cos\alpha + \frac{d\gamma}{dt}\sin\alpha\sin\beta\right) \\
& + j\left(\frac{d\beta}{dt}\sin\alpha - \frac{d\gamma}{dt}\cos\alpha\sin\beta\right) \\
& + k\left(\frac{d\alpha}{dt} + \frac{d\gamma}{dt}\cos\beta\right).
\end{aligned}$$

Chapter 3

S3-1

$$a_0 + i'a_3 = 1, \qquad a_0 - i'a_3 = 0,$$
$$-i'a_1 + a_2 = 0, \qquad -i'a_1 - a_2 = 0,$$

hence, $a_0 = \frac{1}{2}$, $a_3 = \frac{-i'}{2}$, $a_2 = 0 = a_1$, or

$$e_1 = \frac{1}{2}(1 - i'k); \qquad e_1^2 = e_1;$$

similarly

$$e_2 = \frac{1}{2}(1 + i'k); \qquad e_2^2 = e_2.$$

$$a = (a_0 + i'a_0', a_1 + i'a_1', a_2 + i'a_2', a_3 + i'a_3'),$$

$$u = ae_1 = \frac{1}{2}\begin{bmatrix} a_0 + i'a_3 + i'a_0' - a_3', \\ a_1 - i'a_2 + i'a_1' + a_2', \\ a_1 - i'a_2 + i'a_1' + a_2', \\ -i'a_0 + a_3 + a_0' + i'a_3' \end{bmatrix},$$

$$v = e_1a = \frac{1}{2}\begin{bmatrix} a_0 + i'a_3 + i'a_0' - a_3', \\ a_1 + i'a_2 + i'a_1' - a_2', \\ -i'a_1 + a_2 + a_1' + i'a_2', \\ -i'a_0 + a_3 + a_0' + i'a_3' \end{bmatrix}.$$

S3-2

$$x + y = 3 + i'i + (2 + i')\,j + (1 + 3i')\,k,$$
$$xy = (2 - 3i') + (-3 + 8i')\,i + (7 + 2i')\,j + (2 + 3i')\,k,$$
$$yx = (2 - 3i') + (3 - 4i')\,i + (1 + 2i')\,j + (2 + 3i')\,k,$$

$$x^2 = (-2 - 4i') + 2i'i + (4 + 2i')\,j + 2k, \qquad y^2 = 13 + 12\,i'k,$$
$$xx_c = 4 + 4i', \qquad yy_c = -5,$$

$$x^{-1} = \frac{x_c}{xx_c} = \frac{1}{8}\,(1 - i', -1 - i', -3 + i', -1 + i'),$$
$$y^{-1} = \frac{1}{5}\,(-2 + 3i'k),$$
$$y^{-1}x^{-1} = \frac{1}{40}\,(1 + 5i', 5 + 11i', 9 - 5i', 5 + i') = (xy)^{-1}.$$

S3-3

$$r_1 = \pm\frac{a + a^*}{|1 + d|} = \pm\frac{1}{2}\left(1 + \sqrt{3}k\right),$$
$$b_1 = \pm\frac{1 + d}{|1 + d|} = \pm\frac{1}{2}\left(3 + i'i\sqrt{5}\right),$$
$$d = aa_c^* = \frac{1}{2}\left(7 + 3i'i\sqrt{5}\right),$$

$$r_2 = \pm\frac{a + a^*}{|1 + d'|} = \pm\frac{1}{2}\left(1 + \sqrt{3}k\right),$$
$$b_2 = \pm\frac{1 + d'}{|1 + d'|} = \pm\frac{1}{4}\left(6 - i'\sqrt{5} - i'j\sqrt{15}\right),$$
$$d' = a_c^*a = \frac{1}{4}\left(14 - 3i'i\sqrt{5}i - 3i'j\sqrt{15}\right).$$

S3-4

$$x \cdot y = \frac{1}{2}(xy_c + yx_c) = 2,$$

$$B = x \wedge y = \frac{1}{2}(xy_c - yx_c) = (1 + 2i')\, i + (-1 + 2i')\, j - i'k,$$

$$B' = z \wedge w = (1 - 9i')\, i - 3i'j + (-3 - 3i')\, k,$$

$$F = y \wedge z = -i - 2i'j + 3i'k,$$

$$T = x \wedge (y \wedge z) = \frac{1}{2}(xF^* + Fx) = i' + 2i - 3j - 2k,$$
$$T' = (x \wedge y) \wedge z = z \wedge (x \wedge y) = i' + 2i - 3j - 2k,$$

$$B \cdot z = -z \cdot B = -\frac{1}{2}(zB^* - Bz) = 2 + 6i'i + 6i'j - 2i'k \in V,$$

$$w \cdot T = \frac{1}{2}(wT^* + Tw_c) = (3 + i')\, i + (1 + 8i')\, j + (1 - 11i')\, k \in B,$$

$$B \cdot B' = S\left(\frac{BB' + B'B}{2}\right) = -22,$$
$$B \wedge B' = P\left(\frac{BB' + B'B}{2}\right) = i',$$
$$B \cdot T = \frac{BT + TB^*}{2} = -5 - 6i'i + i'j - 10\, i'k \in V.$$

S3-5

$$X = ct + xi'i + yi'j + zi'k,$$
$$X' = ct' + x'i'i + y'i'j + z'i'k,$$

$$b = \left(\cosh\frac{\varphi}{2}, i' \sinh\frac{\varphi}{2}, 0, 0\right),$$
$$\gamma = \frac{1}{\sqrt{1 - \frac{v^2}{c^2}}} = \cosh\varphi = \frac{7}{2},$$

$$\cosh^2\frac{\varphi}{2} = \frac{\cosh\varphi + 1}{2}, \qquad \sinh^2\frac{\varphi}{2} = \frac{\cosh\varphi - 1}{2},$$

$$\cosh\frac{\varphi}{2} = \frac{3}{2}, \qquad \sinh\frac{\varphi}{2} = \frac{\sqrt{5}}{2}, \qquad b = \left(\frac{3}{2}, i'\frac{\sqrt{5}}{2}, 0, 0\right),$$

$$\gamma' = \frac{1}{\sqrt{1-\frac{v'^2}{c^2}}} = \cosh\varphi' = \sqrt{\frac{7}{6}}, \qquad \sinh\varphi' = \sqrt{\frac{1}{6}},$$

$$\begin{aligned} X' &= (0, i', i', 0), \\ V' &= \left(\sqrt{\frac{7}{6}}, 0, i'\sqrt{\frac{1}{6}}, 0\right), \\ X &= bX'b_c^*, \qquad V = bV'b_c^*, \end{aligned}$$

$$\begin{aligned} V &= \left(\frac{7}{2}\sqrt{\frac{7}{6}}, i'\frac{1}{2}\sqrt{\frac{105}{2}}, i'\sqrt{\frac{1}{6}}, 0\right), \\ F' &= -\mathbf{B}' + i'\frac{\mathbf{E}'}{c} = \left(0, i'\frac{E'_x}{c}, i'\frac{E'_y}{c}, i'\frac{E'_z}{c}\right), \\ F &= bF'b_c \end{aligned}$$

$$\begin{aligned} E_x &= E'_x, \qquad & E_y &= \frac{7}{2}E'_y, \qquad & E_z &= \frac{7}{2}E'_z, \\ B_x &= 0, & B_y &= -\frac{3\sqrt{5}}{2c}E'_z, & B_z &= \frac{3\sqrt{5}}{2c}E'_y, \end{aligned}$$

$$F = \left(0, -\mathbf{B} + i'\frac{\mathbf{E}}{c}\right).$$

S3-6

$$A\,(0, i', -i', 0), \quad B\,(0, i', i', 0), \quad C\,(0, -i', -i', 0), \quad D\,(0, -i', -i', 0),$$

$$\begin{aligned} &A'\left(0, \frac{i'}{3}, -\frac{i'}{3}, -\frac{2i'}{3}\right), \qquad && B'\left(0, \frac{i'}{3}, \frac{i'}{3}, -\frac{2i'}{3}\right), \\ &C'\left(0, -\frac{i'}{3}, \frac{i'}{3}, -\frac{2i'}{3}\right), && D'\left(0, -\frac{i'}{3}, -\frac{i'}{3}, -\frac{2i'}{3}\right). \end{aligned}$$

Chapter 4

S4-1

$$A_c = -I - 2J + iK, \qquad B_c = -j - kI - 2kK,$$

$$\begin{aligned} A + B &= j + I + 2J - iK + kI + 2kK, \\ A - B &= I + 2J - iK - j - kI - 2kK, \end{aligned}$$

$$AA_c = 4, \qquad BB_c = -4,$$

$$A^{-1} = \frac{A_c}{AA_c} = \frac{1}{4}(-I - 2J + iK),, \qquad B^{-1} = \frac{1}{4}(j + kI + 2kK),$$

$$\begin{aligned} AB &= -2j - k + jI + 3jJ + 4kI - 2kJ - 3kK, \\ BA &= -2j - k + jI + 3jJ - 4kI + 2kJ + 3kK, \end{aligned}$$

$$\begin{aligned} (AB)_c &= 2j - k + jI + 3jJ - 4kI + 2kJ + 3kK, \\ (BA)_c &= -2j - k + jI + 3jJ + 4kI - 2kJ - 3kK, \end{aligned}$$

$$(AB)(AB)_c = (AB)_c(AB) = -16, \qquad (BA)(BA)_c = -16,$$

$$\begin{aligned} (AB)^{-1} &= \frac{(AB)_c}{(AB)(AB)_c} \\ &= \frac{1}{16}(-2j + k - jI + 4kI - 3jJ - 2kJ - 3kK) = B^{-1}A^{-1} \\ (BA)^{-1} &= \frac{1}{16}(2j + k - jI - 4kJ - 3jJ + 2kJ + 3kK), \end{aligned}$$

$$[A, B] = \frac{1}{2}(AB - BA) = -2j + 4kI - 2kJ - 3kK.$$

S4-2

$$x \cdot y = -\frac{1}{2}(xy + yx) = 2,$$

$$\begin{aligned} B &= x \wedge y = -\frac{1}{2}(xy - yx) = I - J + 2iI + 2iJ - iK, \\ B' &= z \wedge w = I - 3K - 9iI - 3iJ - 3iK, \end{aligned}$$

$$F = y \wedge z = -I - 2iJ + 3iK,$$

$$\begin{aligned} T &= x \wedge (y \wedge z) = x \wedge F = \frac{1}{2}(xF + Fx) \\ &= k + 2jI - 3jJ - 2jK, \\ T' &= (x \wedge y) \wedge z = z \wedge (x \wedge y) = \frac{1}{2}(zB + Bz) = T \\ &= k + 2jI - 3jJ - 2jK, \end{aligned}$$

$$B \cdot z = -z \cdot B = -\frac{1}{2}(zB - Bz) = 2j + 6kI + -6kJ - 2kK,$$

$$w \cdot T = -\frac{1}{2}(wT + Tw) = 3I + J + K + iI + 8iJ - 11\,iK,$$

$$B \cdot B' = S\left[\frac{1}{2}(BB' + B'B)\right] = -22,$$
$$B \wedge B' = PS\left[\frac{1}{2}(BB' + B'B)\right] = i,$$
$$B \cdot T = \frac{1}{2}(BT + TB) = -5j - 6kI + kJ - 10\,kK.$$

S4-3

$$S_1 = -\frac{1}{2}(xy - yx) = 3I - J - 5K,$$
$$S_2 = -\frac{1}{2}(zw - wz) = 9I - 3J - 3K,$$

$$V = S_1 \wedge z = z \wedge S_1 = \frac{1}{2}(zS_1 + S_1 z) = 10\,k,$$
$$S_1 \wedge S_2 = 0,$$

$$w = w_{\|} + w_{\perp},$$
$$w_{\|} = (w \cdot S_1)\, S_1^{-1} = S_1^{-1}(S_1 \cdot w) = \frac{169}{50}kI - \frac{21}{25}kJ + \frac{27}{10}kK,$$
$$w_{\perp} = (w \wedge S_1)\, S_1^{-1} = S_1^{-1}(S_1 \wedge w) = -\frac{69}{50}kI + \frac{46}{25}kJ + \frac{23}{10}kK,$$

$$S_1 = S_{1\|} + S_{1\perp},$$
$$S_{1\|} = (S_1.S_2)\, S_2^{-1} = -\frac{54}{11}I - \frac{18}{11}J - \frac{18}{11}K,$$
$$S_{1\perp} = \Big\{(S_1 \wedge S_2) + [S_1, S_2]\Big\} S_2^{-1} = -\frac{21}{11}I - \frac{26}{11}J - \frac{37}{11}K.$$

Chapter 5

S5-1

$$\gamma = \frac{1}{\sqrt{1 - \frac{v^2}{c^2}}} = \cosh\varphi = \frac{3}{2\sqrt{2}},$$

$$\cosh^2\frac{\varphi}{2} = \frac{\cosh\varphi + 1}{2}, \qquad \sinh^2\frac{\varphi}{2} = \frac{\cosh\varphi - 1}{2},$$
$$\cosh\frac{\varphi}{2} = \frac{1}{2}\sqrt{\frac{3}{\sqrt{2}} + 2} = \alpha, \qquad \sinh\frac{\varphi}{2} = \frac{1}{2}\sqrt{\frac{3}{\sqrt{2}} - 2} = \beta,$$

$$b = \frac{1}{2}\left(\sqrt{\frac{3}{\sqrt{2}} + 2} + iI\sqrt{\frac{3}{\sqrt{2}} - 2}\right),$$

$$r = \cos\frac{\theta}{2} + I\sin\frac{\theta}{2} = \cos\frac{\pi}{8} + I\sin\frac{\pi}{8},$$

$$a = rb = \alpha\cos\frac{\pi}{8} - \beta\sin\frac{\pi}{8}i + \alpha\sin\frac{\pi}{8} + \alpha\sin\frac{\pi}{8}I + \beta\cos\frac{\pi}{8}iI,$$

$$\begin{aligned} X &= aX'a_c, \\ X &= ctj + kIx + kJy + kKz, \\ X' &= ct'j + kIx' + kJy' + kKz'. \end{aligned}$$

S5-2

$$\begin{aligned} b_1 &= \frac{1}{2}\left(\sqrt{\frac{3}{\sqrt{2}} + 2} + iI\sqrt{\frac{3}{\sqrt{2}} - 2}\right), \\ b_2 &= \frac{1}{2}\left(\sqrt{\frac{3}{\sqrt{2}} + 2} + iJ\sqrt{\frac{3}{\sqrt{2}} - 2}\right), \end{aligned}$$

$$\begin{aligned} a &= b_2b_1 = \frac{1}{4}\left[\left(2 + \frac{3}{\sqrt{2}}\right) + \frac{1}{\sqrt{2}}iI + \frac{1}{\sqrt{2}}iJ + \left(-2 + \frac{3}{\sqrt{2}}\right)K\right], \\ a &= br, \end{aligned}$$

$$\begin{aligned} b &= \pm\frac{1+d}{\sqrt{2+d+d_c}}, d = a\overline{a}_c = \frac{9}{8} + \frac{1}{2\sqrt{2}}iI + \frac{3}{8}iJ, \\ b &= \pm\left[\frac{\sqrt{17}}{4} + \frac{1}{\sqrt{34}}iI + \frac{3}{4\sqrt{17}}iJ\right], \end{aligned}$$

$$r = \pm\frac{a+\overline{a}}{\sqrt{2+d+d_c}} = \pm\left[\left(2 + \frac{3}{\sqrt{2}}\right)\frac{1}{\sqrt{17}} + \left(-2 + \frac{3}{\sqrt{2}}\right)\frac{1}{\sqrt{17}}K\right],$$

$$\cos\frac{\theta}{2} = \left(2 + \frac{3}{\sqrt{2}}\right)\frac{1}{\sqrt{17}}, \qquad \theta = 3,37^\circ,$$

$$\cosh\frac{\varphi}{2} = \frac{\sqrt{17}}{4}, \qquad \sinh\frac{\varphi}{2} = \frac{1}{4}, \qquad \varphi = 0,494,$$

$$b = \pm\left[\frac{\sqrt{17}}{4} + \frac{1}{4\sqrt{34}}4iI + \frac{3}{\sqrt{17}}iJ\right],$$

$$\text{direction } \mathbf{u}\left(\frac{1}{\sqrt{34}}4, \frac{3}{\sqrt{17}}, 0\right), \qquad \tanh\varphi = 0,4581 = \frac{v}{c}.$$

S5-3

$$x'_A = (0,0,0)\,, \qquad x'_B = \left(-\frac{1}{2}, 0, -\frac{1}{2}\right),$$
$$x'_C = \left(-\frac{3}{7}, \frac{1}{7}, -\frac{2}{7}\right), \qquad x'_D = \left(-\frac{1}{3}, \frac{1}{6}, -\frac{1}{6}\right),$$
$$x'_E = \left(-\frac{4}{9}, \frac{1}{9}, -\frac{1}{9}\right), \qquad x'_F = \left(-\frac{1}{2}, 0, 0\right),$$
$$x'_G = \left(-\frac{3}{5}, 0, -\frac{1}{5}\right), \qquad x'_H = \left(-\frac{1}{2}, \frac{1}{10}, -\frac{1}{5}\right).$$

S5-4

$$AB = \begin{bmatrix} 1-k & -1+k \\ -1+k & -1+k \end{bmatrix}, \qquad BA = \begin{bmatrix} 0 & j \\ i & 0 \end{bmatrix},$$

$$A^{-1} = \frac{1}{2}\begin{bmatrix} 1 & -k \\ -j & -i \end{bmatrix}, \qquad B^{-1} = \frac{1}{2}\begin{bmatrix} 1 & -i \\ -k & -j \end{bmatrix},$$

$$(AB)^{-1} = \frac{1}{4}\begin{bmatrix} 1+k & -1-k \\ -1-k & -1-k \end{bmatrix}, \qquad (BA)^{-1} = \frac{1}{2}\begin{bmatrix} 0 & -i \\ -j & 0 \end{bmatrix}.$$

There is no inverse of C.

Chapter 6

S6-1 Velocity of B in the reference frame at rest

$$v_x = -v, \qquad v_y = 0, \qquad v_z = 0$$

velocity of B with respect to A

$$v'_x = \frac{v_x - w}{1 - \frac{v_x w}{c^2}} = -0,975\,c, \qquad w = v = 0,8\,c,$$
$$v'_y = 0, \qquad v'_z = 0;$$

The relative velocity remains smaller than c.

S6-2

$$L = \sum X_i \wedge P_i = \sum (\mathbf{r}_i \times \mathbf{p}_i) + i \sum \left(\mathbf{r}_i \frac{E_i}{c} - ct\mathbf{p}_i \right) = \text{const.}$$

hence

$$(\mathbf{r}_i \times \mathbf{p}_i) = \mathbf{C}_1 , \qquad \sum \left(\mathbf{r}_i \frac{E_i}{c} - ct\mathbf{p}_i \right) = \mathbf{C}_2,$$

$$\mathbf{R} = \frac{\sum E_i \mathbf{r}_i}{\sum E_i},$$

$$\frac{\sum E_i \mathbf{r}_i}{\sum E_i} - tc^2 \frac{\sum \mathbf{p}_i}{\sum E_i} = \mathbf{C}_3,$$

$$\mathbf{R} = \mathbf{C}_3 + \mathbf{v}t, \qquad \mathbf{v} = c^2 \frac{\sum \mathbf{p}_i}{\sum E_i} = \mathbf{C}_4.$$

S6-3 In the proper frame (instantaneous, Galilean), the four-acceleration is

$$A' = kIg$$

with the invariant

$$AA_c = -g^2;$$

In the frame at rest K, the four-velocity and the four-acceleration are respectively

$$u = c\,(j \cos h\varphi + kI \sinh \varphi)\,, \qquad \tanh \varphi = \frac{v}{c},$$

$$A = c \frac{d\varphi}{d\tau} (j \sinh \varphi + kI \cosh \varphi)\,, \qquad \tau : \text{proper time}$$

with

$$AA_c = -g^2 = c^2 \left(\frac{d\varphi}{d\tau} \right)^2 (\sinh^2 \varphi - \cosh^2 \varphi)$$
$$= -c^2 \left(\frac{d\varphi}{d\tau} \right)^2$$

hence

$$\frac{d\varphi}{d\tau} = \frac{g}{c}, \qquad \varphi = \frac{g\tau}{c}$$

and

$$u = c \left(j \cos h \frac{g\tau}{c} + kI \sinh \frac{g\tau}{c} \right) = \frac{dX}{d\tau},$$

$$X = \frac{c^2}{g} \left(j \sin h \frac{g\tau}{c} + kI \cosh \frac{g\tau}{c} \right) = jct + kIx;$$

finally, one has

$$t = \frac{c}{g}\sinh\varphi, \qquad x = \frac{c^2}{g}(\cosh\varphi - 1)$$

and

$$\left(x + \frac{c^2}{g}\right)^2 - (ct)^2 = \frac{c^4}{g^2}.$$

For $\varphi \ll 1$

$$t \simeq \frac{c}{g}\varphi = \tau, \qquad x = \frac{c^2}{g}\frac{\varphi^2}{2} = \frac{1}{2}g\tau^2 = \frac{1}{2}gt^2.$$

Chapter 7

S7-1

$$A = bA'b_c \qquad \left(b = \cosh\frac{\varphi}{2} + iI\sinh\frac{\varphi}{2},\ \tanh\varphi = \frac{v}{c}\right),$$

$$A' = j\frac{V'}{c} + k\mathbf{A}' = j\frac{V'}{c} \qquad \left(\mathbf{A}' = 0,\ V' = \frac{q}{4\pi\varepsilon_0 r'}\right),$$

$$A = j\frac{V}{c} + k\mathbf{A},$$

$$r' = \left(x'^2 + y'^2 + z'^2\right)^{1/2}, \qquad x' = \gamma(x - vt), \qquad y' = y, \qquad z' = z,$$

$$V = \gamma V' = \frac{q}{4\pi\varepsilon_0}\frac{1}{\left[(x - vt)^2 + \frac{y^2+z^2}{\gamma^2}\right]^{1/2}}, \qquad \mathbf{A} = \gamma\frac{\mathbf{v}}{c^2}V' = \frac{\mathbf{v}}{c^2}V,$$

(potentials of Lienard and Wiechert of an electric charge in rectilinear motion).

$$F' = -\mathbf{B}' + \frac{\mathbf{E}'}{c}, \qquad F = -\mathbf{B} + \frac{\mathbf{E}}{c},$$

$$\mathbf{B}' = 0, \qquad \mathbf{E}' = \frac{q}{4\pi\varepsilon_0}\frac{\mathbf{r}'}{r'^3},$$

$$F = bF'b_c,$$

$$B_x = 0,$$

$$B_y = -\frac{q}{4\pi\varepsilon_0}\frac{\gamma v z'}{cr'^3} = -\gamma E'_z\frac{v}{c} = -E_z\frac{v}{c},$$

$$B_z = \frac{q}{4\pi\varepsilon_0}\frac{\gamma v y'}{cr'^3} = \gamma E'_y\frac{v}{c} = E_y\frac{v}{c},$$

$$E_x = E'_x = \frac{q}{4\pi\varepsilon_0}\frac{x'}{r'^3}, \qquad E_y = \gamma E'_y, \qquad E_z = \gamma E'_z,$$

$$x' = \gamma(x - vt), \qquad y' = y, \qquad z' = z,$$

$$E_x = \frac{q}{4\pi\varepsilon_0}\frac{(x-vt)}{\gamma^2 r^{*3/2}}, \quad r^* = \left[(x-vt)^2 + \frac{y^2+z^2}{\gamma^2}\right]^{1/2},$$
$$E_y = \frac{q}{4\pi\varepsilon_0}\frac{y}{\gamma^2 r^{*3/2}},$$
$$E_z = \frac{q}{4\pi\varepsilon_0}\frac{z}{\gamma^2 r^{*3/2}}.$$

S7-2

$$F' = -\mathbf{B}' + \frac{\mathbf{E}'}{c},$$
$$\mathbf{B}' = 0, \qquad \mathbf{E}' = \frac{\lambda_0}{2\pi\varepsilon_0}\frac{\mathbf{n}}{r'},$$

$$E'_2 = \frac{\lambda_0}{2\pi\varepsilon_0 r'}, \qquad E'_1 = E'_3 = 0,$$
$$F = bF'b_c,$$
$$b = \cosh\frac{\varphi}{2} + iI\sinh\frac{\varphi}{2}, \qquad \tanh\varphi = \frac{v}{c},$$
$$B_3 = \frac{\lambda_0}{2\pi\varepsilon_0 r}\frac{\gamma v}{c^2} = \frac{\mu_0\lambda v}{2\pi r} = \frac{\mu_0 i}{2\pi r},$$
$$E_3 = \frac{\lambda_0\gamma}{2\pi\varepsilon_0 r} = \frac{\lambda}{2\pi\varepsilon_0 r},$$

in K; the linear density of mobile charges is $\lambda_K = \gamma\lambda_0 = \lambda$ (by hypothesis, $\lambda_K - \lambda = 0$) and $i = \lambda v$ ($r = r'$). At the point M in K, the total electric field is nil

$$\mathbf{E}_T = 0, \qquad \mathbf{B}_T = \frac{\mu_0 i}{2\pi r}\mathbf{e}_z.$$

In K,

$$\mathbf{f} = q\mathbf{v}\times\mathbf{B},$$
$$\mathbf{f}(M) = -qvB\mathbf{e}_y;$$

in the proper frame, the electromagnetic field is

$$F'' = -\mathbf{B}'' + \frac{\mathbf{E}''}{c} = b_c F_T b$$

with

$$F_T = -\mathbf{B}, \qquad \mathbf{B} = \frac{\mu_0 i}{2\pi r}\mathbf{e}_z,$$

hence

$$\mathbf{B}''=\frac{\mu_0 i}{2\pi r}\gamma\mathbf{e}_z,$$
$$\mathbf{E}''=-\frac{\mu_0 i}{2\pi r}\gamma v\mathbf{e}_y;$$

the particle being at rest in the proper frame,

$$\mathbf{f}''=qE''=-\frac{q\mu_0 i}{2\pi r}\gamma v\mathbf{e}=\gamma\mathbf{f}.$$

S7-3

$$F'=b_cFb,\qquad b=\left(\cosh\frac{\varphi}{2}+\frac{iI}{\sqrt{2}}\sinh\frac{\varphi}{2}+\frac{iJ}{\sqrt{2}}\sinh\frac{\varphi}{2}\right),$$

$$\tanh\varphi=\frac{v}{c}=\frac{1}{2},\qquad \cosh\varphi=\gamma=\frac{1}{\sqrt{1-\frac{v^2}{c^2}}}=\frac{2}{\sqrt{3}},$$

$$\cosh^2\frac{\varphi}{2}=\frac{\cosh\varphi+1}{2}=\frac{\frac{2}{\sqrt{3}}+1}{2},\qquad \sinh^2\frac{\varphi}{2}=\frac{\cosh\varphi-1}{2}=\frac{\frac{2}{\sqrt{3}}-1}{2},$$

$$\cosh\frac{\varphi}{2}=\sqrt{\frac{\frac{2}{\sqrt{3}}+1}{2}},\qquad \sinh\frac{\varphi}{2}=\sqrt{\frac{\frac{2}{\sqrt{3}}-1}{2}},$$

$$B_1'=\frac{1}{2}+\frac{1}{\sqrt{3}},\qquad E_1'=1-\frac{2}{\sqrt{3}},$$
$$B_2'=\frac{1}{2}-\frac{1}{\sqrt{3}},\qquad E_2'=1+\frac{2}{\sqrt{3}},$$
$$B_3'=-\frac{1}{c}\sqrt{\frac{2}{3}},\qquad E_1'=-\frac{c}{\sqrt{6}}.$$

S7-4

$$\nabla F=0,$$
$$i\nabla F=-\nabla(iF)=0,$$
$$\nabla F^*=0.$$

Appendix B

Formulary: multivector products within $\mathbb{H}(\mathbb{C})$

Let x, y be four-vectors, B, B' bivectors, T, T' trivectors and P, P' pseudoscalars

$$\begin{aligned}
&x = x^0 + i'x^1 + i'x^2 + i'x^3,\\
&y = y^0 + i'y^1 + i'y^2 + i'y^3,\\
&B = \mathbf{a} + i'\mathbf{b}, \qquad B' = \mathbf{a}' + i'\mathbf{b}',\\
&(\mathbf{a} = a^1 i + a^2 j + a^3 k,\ \mathbf{b} = b^1 i + b^2 j + b^3 k),\\
&T = i't^0 + \mathbf{t}, \qquad T' = i't'^0 + \mathbf{t}'(\mathbf{t} = t^1 i + t^2 j + t^3 k),\\
&P = i's^0, \qquad P' = i's'^0.
\end{aligned}$$

$S\,[A]$, $P\,[A]$ designate respectively the scalar and pseudoscalar parts of the complex quaternion A. For any two arbitrary boldface quantities $(\mathbf{x}, \mathbf{y}, \mathbf{a}, \mathbf{b}, \mathbf{t})$ the following abridged notation is used:

$$\begin{aligned}
\mathbf{x} \cdot \mathbf{y} &= x^1 y^1 + x^2 y^2 + x^3 y^3,\\
\mathbf{x} \times \mathbf{y} &= (x^2 y^3 - x^3 y^2) i + (x^3 y^1 - x^1 y^3) j + (x^1 y^2 - x^2 y^1) k.
\end{aligned}$$

Products with four-vectors

$$\begin{aligned}
x \cdot y &= \frac{1}{2}\,(x y_c + y x_c) \in S\\
&= x^0 y^0 - x^1 y^1 - x^2 y^2 - x^3 y^3\\
&= y \cdot x,
\end{aligned}$$

$$\begin{aligned} x \wedge y &= \frac{1}{2}(xy_c - yx_c) \in B \\ &= \mathbf{x} \times \mathbf{y} + i'\left(y^0\mathbf{x} - x^0\mathbf{y}\right) \\ &= -y \wedge x. \end{aligned}$$

Products with bivectors

$$\begin{aligned} x \cdot B &= \frac{1}{2}(xB^* - Bx) \in V \\ &= -\mathbf{x} \cdot \mathbf{b} + i'\left(-x^0\mathbf{b} + \mathbf{x} \times \mathbf{a}\right) \\ &\equiv -B \cdot x, \end{aligned}$$

$$\begin{aligned} x \wedge B &= \frac{1}{2}(xB^* + Bx) \in T \\ &= -i'\mathbf{x} \cdot \mathbf{a} + \left(x^0\mathbf{a} + \mathbf{x} \times \mathbf{b}\right) \\ &\equiv B \wedge x. \end{aligned}$$

$$\begin{aligned} B \cdot B' &= S\left[\frac{1}{2}(BB' + B'B)\right] \\ &= -\mathbf{a} \cdot \mathbf{a}' + \mathbf{b} \cdot \mathbf{b}' \\ &\equiv B' \cdot B, \end{aligned}$$

$$\begin{aligned} B \wedge B' &= P\left[\frac{1}{2}(BB' + B'B)\right] \\ &= -i'(\mathbf{b} \cdot \mathbf{a}' + \mathbf{a} \cdot \mathbf{b}') \\ &\equiv B' \wedge B. \end{aligned}$$

Products with trivectors

$$\begin{aligned} x \cdot T &= \frac{1}{2}(xT^* + Tx_c) \in B \\ &= x^0\mathbf{t} + t^0\mathbf{x} + i'(\mathbf{x} \times \mathbf{t}) \\ &\equiv T \cdot x \end{aligned}$$

$$\begin{aligned} x \wedge T &= \frac{1}{2}(xT^* - Tx_c) \in P \\ &= i'\left(-x^0t^0 - \mathbf{x} \cdot \mathbf{t}\right) \\ &\equiv -T \wedge x, \end{aligned}$$

$$
\begin{aligned}
B \cdot T &= -\frac{1}{2}\left(BT + TB^*\right) \in V \\
&= (\mathbf{a} \cdot \mathbf{t}) - i'\left(t^0\mathbf{a} + \mathbf{b} \times \mathbf{t}\right) \\
&\equiv T \cdot B, \\
T \cdot T' &= \frac{1}{2}\left(TT'^* + T'T^*\right) \in S \\
&= t^0 t'^0 - \mathbf{t} \cdot \mathbf{t}' \\
&\equiv T' \cdot T.
\end{aligned}
$$

Products with pseudoscalars

$$
\begin{aligned}
x \cdot P &= \frac{1}{2}\left(xP^* - Px\right) \in T \\
&= -i's^0x^0 + s^0\mathbf{x} \\
&\equiv -P \cdot x,
\end{aligned}
$$

$$
\begin{aligned}
B \cdot P &= \frac{1}{2}\left(BP + PB\right) \in B \\
&= -s^0\mathbf{b} + i's^0\mathbf{a} \\
&\equiv P \cdot B,
\end{aligned}
$$

$$
\begin{aligned}
T \cdot P &= -\frac{1}{2}\left(TP^* - PT\right) \in V \\
&= -s^0t^0 + i's^0\mathbf{t} \\
&\equiv -P \cdot T,
\end{aligned}
$$

$$
\begin{aligned}
P.P' &= S\left[\frac{1}{2}\left(PP' + P'P\right)\right] \in S \\
&= -s^0 s'^0 \\
&\equiv P' \cdot P.
\end{aligned}
$$

Appendix C

Formulary: multivector products within $\mathbb{H} \otimes \mathbb{H}$ (over $\mathbb{R}$)

Let x, y be four-vectors, B, B' bivectors, T, T' trivectors and P, P' pseudoscalars

$$
\begin{aligned}
&x = jx^0 + k\mathbf{x} \qquad (\mathbf{x} = x^1 I + x^2 J + x^3 K),\\
&y = jy^0 + k\mathbf{y},\\
&B = \mathbf{a} + i\mathbf{b}, \qquad B' = \mathbf{a}' + i\mathbf{b}',\\
&(\mathbf{a} = a^1 I + a^2 J + a^3 K,\ \mathbf{b} = b^1 I + b^2 J + b^3 K),\\
&T = kt^0 + j\mathbf{t}, \qquad T' = kt'^0 + j\mathbf{t}'\ (\mathbf{t} = t^1 I + t^2 J + t^3 K),\\
&P = is^0, \qquad P' = is'^0.
\end{aligned}
$$

$S[A]$, $P[A]$ designate respectively the scalar and pseudoscalar parts of the Clifford number A. For any two arbitrary boldface quantities $(\mathbf{x}, \mathbf{y}, \mathbf{a}, \mathbf{b}, \mathbf{t})$ the following abridged notation is used:

$$
\begin{aligned}
\mathbf{x} \cdot \mathbf{y} &= x^1 y^1 + x^2 y^2 + x^3 y^3,\\
\mathbf{x} \times \mathbf{y} &= (x^2 y^3 - x^3 y^2) I + (x^3 y^1 - x^1 y^3) J + (x^1 y^2 - x^2 y^1) K.
\end{aligned}
$$

Products with four-vectors

$$\begin{aligned} x \cdot y &= -\frac{1}{2}(xy + yx) \in S \\ &= x^0y^0 - x^1y^1 - x^2y^2 - x^3y^3 \\ &= y \cdot x, \\ x \wedge y &= -\frac{1}{2}(xy - yx) \in B \\ &= \mathbf{x} \times \mathbf{y} + i\left(y^0\mathbf{x} - x^0\mathbf{y}\right) \\ &= -y \wedge x. \end{aligned}$$

Products with bivectors

$$\begin{aligned} x \cdot B &= \frac{1}{2}(xB - Bx) \in V \\ &= -j\mathbf{x} \cdot \mathbf{b} + k\left(-x^0\mathbf{b} + \mathbf{x} \times \mathbf{a}\right) \\ &= -B \cdot x, \\ x \wedge B &= \frac{1}{2}(xB + Bx) \in T \\ &= -k\mathbf{x} \cdot \mathbf{a} + j\left(x^0\mathbf{a} + \mathbf{x} \times \mathbf{b}\right) \\ &\equiv B \wedge x. \end{aligned}$$

$$\begin{aligned} B \cdot B' &= S\left[\frac{1}{2}(BB' + B'B)\right] \\ &= -\mathbf{a} \cdot \mathbf{a}' + \mathbf{b} \cdot \mathbf{b}' \\ &= B' \cdot B, \\ B \wedge B' &= P\left[\frac{1}{2}(BB' + B'B)\right] \\ &= -i\left(\mathbf{b} \cdot \mathbf{a}' + \mathbf{a} \cdot \mathbf{b}'\right) \\ &\equiv B' \wedge B. \end{aligned}$$

Products with trivectors

$$
\begin{aligned}
x \cdot T &= -\frac{1}{2}(xT + Tx) \in B \\
&= x^0\mathbf{t} + t^0\mathbf{x} + i(\mathbf{x} \times \mathbf{t}) \\
&\equiv T \cdot x, \\
x \wedge T &= -\frac{1}{2}(xT - Tx) \in P \\
&= i\left(-x^0 t^0 - \mathbf{x} \cdot \mathbf{t}\right) \\
&\equiv -T \wedge x,
\end{aligned}
$$

$$
\begin{aligned}
B \cdot T &= \frac{1}{2}(BT + TB) \in V \\
&= j(\mathbf{a} \cdot \mathbf{t}) - k\left(t^0\mathbf{a} + \mathbf{b} \times \mathbf{t}\right) \\
&\equiv T \cdot B, \\
T \cdot T' &= -\frac{1}{2}(TT' + T'T) \in S \\
&= t^0 t'^0 - \mathbf{t} \cdot \mathbf{t}' \\
&\equiv T' \cdot T.
\end{aligned}
$$

Products with pseudoscalars

$$
\begin{aligned}
x \cdot P &= \frac{1}{2}(xP - Px) \in T \\
&= -ks^0 x^0 + js^0\mathbf{x} \\
&\equiv -P \cdot x,
\end{aligned}
$$

$$
\begin{aligned}
B \cdot P &= \frac{1}{2}(BP + PB) \in B \\
&= -s^0\mathbf{b} + is^0\mathbf{a} \\
&\equiv P \cdot B,
\end{aligned}
$$

$$
\begin{aligned}
T \cdot P &= \frac{1}{2}(TP - PT) \in V \\
&= -js^0 t^0 + ks^0\mathbf{t} \\
&\equiv -P \cdot T,
\end{aligned}
$$

$$\begin{aligned} P \cdot P' &= S\left[\frac{1}{2}(PP' + P'P)\right] \in S \\ &= -s^0 s'^0 \\ &\equiv P' \cdot P. \end{aligned}$$

Appendix D

Formulary: four-nabla operator ∇ within $\mathbb{H} \otimes \mathbb{H}$ (over $\mathbb{R}$)

Four-nabla operator:

$$\nabla = j\frac{\partial}{c\partial t} - kI\frac{\partial}{\partial x} - kJ\frac{\partial}{\partial y} - kK\frac{\partial}{\partial z},$$

D'Alembertian operator:

$$\Box = \frac{\partial^2}{c^2\partial t^2} - \frac{\partial^2}{\partial x^2} - \frac{\partial^2}{\partial y^2} - \frac{\partial^2}{\partial z}$$

(four-vectors

$$A = jA^0 + kIA^1 + kJA^2 + kKA^3, \qquad B = jB^0 + kIB^1 + kJB^2 + kKB^3;$$

scalars : p, q). Then:

$$\nabla \wedge (\nabla \wedge A) = 0,$$

$$\nabla \wedge (\nabla p) = (\nabla \wedge \nabla)\, p = 0,$$

$$\nabla (\nabla p) = (\nabla \cdot \nabla)\, p = \Box\, p,$$

$$\nabla \cdot (\nabla \wedge A) = \Box\, A - \nabla (\nabla \cdot A)\,,$$

$$\Box\, (\nabla p) = \nabla (\Box\, p)\,,$$

$$\Box\, (\nabla \cdot A) = \nabla \cdot (\Box\, A)\,,$$

$$\Box\, (\nabla \wedge A) = \nabla \wedge (\Box\, A)\,,$$

$$\nabla (pq) = p\nabla q + q\nabla p,$$

$$\Box\, (pq) = p\,\Box\, q + q\,\Box\, p + 2\,(\nabla p) \cdot (\nabla q)\,,$$

$$\nabla \cdot (pA) = p\,(\nabla \cdot A) + A \cdot (\nabla p)\,,$$

$$\nabla \cdot (\nabla \wedge pA) = \Box pA - \nabla (\nabla \cdot pA),$$
$$\nabla \wedge pA = p(\nabla \wedge A) + (\nabla p) \wedge A,$$
$$\Box (A + B) = \Box A + \Box B,$$
$$\nabla \cdot (A + B) = \nabla \cdot A + \nabla \cdot B,$$
$$\nabla \wedge (A + B) = \nabla \wedge A + \nabla \wedge B,$$
$$\Box (p + q) = \Box p + \Box q,$$
$$\nabla (p + q) = \nabla p + \nabla q,$$
$$\nabla \wedge (A \wedge B) = B \wedge (\nabla \wedge A) - A \wedge (\nabla \wedge B),$$
$$\nabla \cdot (A \wedge B) = (\nabla \cdot A) B + (A \cdot \nabla) B - (\nabla \cdot B) A - (B \cdot \nabla) A,$$
$$\nabla (A \cdot B) = -A \cdot (\nabla \wedge B) + (A \cdot \nabla) B - B \cdot (\nabla \wedge A) + (B \cdot \nabla) A.$$

Appendix E

Work-sheet: $\mathbb{H}(\mathbb{C})$ (Mathematica)

```
<<Algebra`Quaternions`
```

(*example $x = x_0+x_1i+x_2j+x_3k = [x_0, x_1, x_2, x_3]$, $x_i \in \mathbb{C}$, I: usual complex imaginary*)

```
x=Quaternion[1,I,2+I,1]

y=Quaternion[2,0,0,3I]

x**y
```

(*the two stars indicate a quaternion product; place the pointer at the end of the last program line and click "Enter" on the numerical pad; result*)

```
out=Quaternion[2-3I,-3+8I,7+2I,2+3I]
```

Appendix F

Work-sheet $\mathbb{H} \otimes \mathbb{H}$ over $\mathbb{R}$ (Mathematica)

```
<<Algebra`Quaternions`
```

(***product of two Clifford numbers**, $a = a_1 + a_2 I + a_3 J + a_4 K$, $b = b_1 + b_2 I + b_3 J + b_4 K$, $a_i, b_i \in \mathbb{H}$*)

```
CP[a_,b_]:=

{(a[[1]]**b[[1]])-(a[[2]]**b[[2]])
  -(a[[3]]**b[[3]])-(a[[4]]**b[[4]]),

(a[[2]]**b[[1]])+(a[[1]]**b[[2]])
  -(a[[4]]**b[[3]])+(a[[3]]**b[[4]]),

(a[[3]]**b[[1]])+(a[[4]]**b[[2]])
  +(a[[1]]**b[[3]])-(a[[2]]**b[[4]]),

(a[[4]]**b[[1]])-(a[[3]]**b[[2]])
  +(a[[2]]**b[[3]])+(a[[1]]**b[[4]])}
```

(***conjugate***)

```
K[a_]:={Quaternion[a[[1,1]],a[[1,2]],-a[[1,3]],a[[1,4]]],

Quaternion[-a[[2,1]],-a[[2,2]],a[[2,3]],-a[[2,4]]],

Quaternion[-a[[3,1]],-a[[3,2]],a[[3,3]],-a[[3,4]]],

Quaternion[-a[[4,1]],-a[[4,2]],a[[4,3]],-a[[4,4]]]}
```

(***sum and difference***)

```
csum[a_,b_]:={a[[1]]+b[[1]],

a[[2]]+b[[2]],a[[3]]+b[[3]],a[[4]]+b[[4]]}

cdif[a_,b_]:={a[[1]]-b[[1]],

a[[2]]-b[[2]],a[[3]]-b[[3]],a[[4]]-b[[4]]}
```

(***multiplication by a scalar** f*)

```
fclif[f_,a_]:={f*a[[1]],f*a[[2]],f*a[[3]],f*a[[4]]}
```

(***products** $\frac{ab+ba}{2}$, $\frac{ab-ba}{2}$*)

```
int[a_,b_]:={fclif[1/2,csum[CP[a,b],CP[b,a]]]}

ext[a_,b_]:={fclif[1/2,cdif[CP[a,b],CP[b,a]]]}
```

(***products** $-\frac{ab+ba}{2}$, $-\frac{ab-ba}{2}$*)

```
mint[a_,b_]:={fclif[-1/2,csum[CP[a,b],CP[b,a]]]}

mext[a_,b_]:={fclif[-1/2,cdif[CP[a,b],CP[b,a]]]}
```

(***example** $A = I + 2J - iK$, $B = j + kI + 2kK$, product $w = AB$*)

```
A:={Quaternion[0,0,0,0],Quaternion[1,0,0,0],

Quaternion[2,0,0,0],Quaternion[0,-1,0,0]}

B:={Quaternion[0,0,1,0],Quaternion[0,0,0,1],

Quaternion[0,0,0,0],Quaternion[0,0,0,2]}

w=CP[A,B]

Simplify[%]
```

(***result** $w = AB = -2j - k + jI + 4kI + 3jJ - 2kJ - 3kK$; the function "Simplify" simplifies numerically or algebraically the result of the line above*)

```
Out={Quaternion[0,0,-2,-1],Quaternion[0,0,1,4],

Quaternion[0,0,3,-2],Quaternion[0,0,0,-3]}
```

Appendix G

Work-sheet: matrices $M_2(\mathbb{H})$ (Mathematica)

```
<<Algebra`Quaternions`
```

(***product of two matrices** $2*2$ over $\mathbb{H}$,

$$a=\begin{bmatrix} a_1 & a_2 \\ a_3 & a_4 \end{bmatrix}, \qquad b=\begin{bmatrix} b_1 & b_2 \\ b_3 & b_4 \end{bmatrix}, \qquad a_i,\ b_i \in \mathbb{H}^*)$$

```
CP[a_,b_]:={(a[[1]]**b[[1]])+(a[[2]]**b[[3]]),

(a[[1]]**b[[2]])+(a[[2]]**b[[4]]),

(a[[3]]**b[[1]])+(a[[4]]**b[[3]]),

(a[[3]]**b[[2]])+(a[[4]]**b[[4]])}
```

(***example**

$$A=\begin{bmatrix} 1 & j \\ k & i \end{bmatrix}, \qquad B=\begin{bmatrix} 1 & k \\ i & j \end{bmatrix} \ *)$$

```
A:={Quaternion[1,0,0,0],Quaternion[0,0,1,0],

Quaternion[0,0,0,1],Quaternion[0,1,0,0]}

B:={Quaternion[1,0,0,0],Quaternion[0,0,0,1],

Quaternion[0,1,0,0],Quaternion[0,0,1,0]}

CP[A,B]
```

(***result**

$$AB = \begin{bmatrix} 1-k & -1+k \\ -1+k & -1+k \end{bmatrix} \text{*)}$$

```
Out={Quaternion[1,0,0,-1],Quaternion[-1,0,0,1],

Quaternion[-1,0,0,1],Quaternion[-1,0,0,1]}
```

Appendix H

Clifford algebras: isomorphisms

General formulas ([18], p. 48)

$M_n(\mathbb{R}), M_n(\mathbb{C}), M_n(\mathbb{H})$: square matrices of order n over $\mathbb{R}$, $\mathbb{C}$, $\mathbb{H}$;

$$M_n(\mathbb{R}) \otimes M_p(\mathbb{R}) \simeq M_{np}(\mathbb{R}),$$
$$M_p(\mathbb{C}) \simeq M_p(\mathbb{R}) \otimes \mathbb{C},$$
$$M_p(\mathbb{H}) \simeq M_p(\mathbb{R}) \otimes \mathbb{H}.$$

Clifford algebra $\mathbb{H}$

$$\boxed{\begin{array}{c} \mathbb{H} = M_1(\mathbb{H}) = M_1(\mathbb{R}) \otimes \mathbb{H}, \\ C^+ = \mathbb{C}. \end{array}}$$

Clifford algebra $\mathbb{H} \otimes \mathbb{H}$

$$\boxed{\begin{array}{c} \mathbb{H} \otimes \mathbb{H} \simeq M_4(\mathbb{R}), \\ C^+ = \mathbb{H} \otimes \mathbb{C} \simeq M_2(\mathbb{C}) \end{array}}$$

Clifford algebra $\mathbb{H} \otimes \mathbb{H} \otimes \mathbb{H}$

$$\boxed{\begin{array}{c} \mathbb{H} \otimes \mathbb{H} \otimes \mathbb{H} \simeq M_4(\mathbb{R}) \otimes \mathbb{H} \simeq M_4(\mathbb{H}), \\ C^+ \simeq \mathbb{H} \otimes \mathbb{H} \otimes \mathbb{C} \simeq M_4(\mathbb{R}) \otimes \mathbb{C} \simeq M_4(\mathbb{C}), \\ \simeq \mathbb{H} \otimes M_2(\mathbb{C}) \simeq M_2\left[\mathbb{H}\left(\mathbb{C}\right)\right] \text{ (Dirac algebra).} \end{array}}$$

Clifford algebra $\mathbb{H} \otimes \mathbb{H} \otimes \mathbb{H} \otimes \mathbb{H}$

$$\boxed{\begin{array}{c} \mathbb{H} \otimes \mathbb{H} \otimes \mathbb{H} \otimes \mathbb{H} \simeq M_4(\mathbb{R}) \otimes M_4(\mathbb{R}) \simeq M_{16}(\mathbb{R}), \\ C^+ \simeq \mathbb{H} \otimes \mathbb{H} \otimes \mathbb{H} \otimes \mathbb{C} \simeq M_4(\mathbb{R}) \otimes \mathbb{H} \otimes C \simeq M_4\left[\mathbb{H}\left(\mathbb{C}\right)\right], \\ \simeq M_4(\mathbb{R}) \otimes M_2(\mathbb{C}) \simeq M_8\left(\mathbb{C}\right). \end{array}}$$

Appendix I

Clifford algebras: synoptic table

$\mathbb{H}(\mathbb{R})$
real quaternions classical vector calculus rotation group SO(3) classical mechanics

$\mathbb{H}(\mathbb{C})$
complex quaternions Lorentz group special relativity classical electromagnetism

$\mathbb{H} \otimes \mathbb{H}(\mathbb{R})$
real Clifford numbers relativistic multivector calculus Lorentz group classical electromagnetism special relativity general relativity

$\mathbb{H} \otimes \mathbb{H}(\mathbb{C})$
complex Clifford numbers Dirac algebra unitary group SU(4) and symplectic unitary group $\mathrm{USp}(2, \mathbb{H})$ relativistic quantum mechanics

Bibliography

[1] S. L. ADLER, *Quaternionic Quantum Mechanics and Quantum Fields*, Oxford University Press, New York, Oxford, 1995.

[2] S. L. ALTMANN, *Rotations, Quaternions, and Double Groups*, Clarendon Press, Oxford, 1986.

[3] J. BASS, *Cours de Mathématiques*, Vol. I, Masson et Cie, Editeurs, Paris, 1964.

[4] J. BATEMAN, The Transformations of the Electrodynamical Equations, *Proceedings of the London Mathematical Society*, Second Series, **8** (1910), 223–264.

[5] W. BAYLIS, ed., *Clifford (Geometric) Algebras with applications to physics, mathematics, and engineering,* Birkhäuser, Boston, 1996.

[6] W. E. BAYLIS, *Electrodynamics: A Modern Geometric Approach*, Birkhäuser, Boston, 1999.

[7] R. BECKER, F. SAUTER, *Theorie der Elektrizität*, Vol. 1, B. G. Teubner, Stuttgart, 1973.

[8] A. BLANCHARD, *Les corps non commutatifs*, Presses Universitaires de France, Collection SUP, 1972.

[9] L. BRAND, *Vector and Tensor Analysis*, Wiley, New York, 1947, p. 412.

[10] E. CARTAN, *Les Systèmes différentiels extérieurs et leurs applications géométriques*, Hermann, Paris, 1971.

[11] E. CARTAN, *Leçons sur la Théorie des Spineurs*, Actualités Scientifiques et Industrielles, n° 643, Hermann, Paris, 1937; I, pp. 12–13.

[12] G. CASANOVA, *L'Algèbre vectorielle*, PUF, Paris, 1976.

[13] W. K. CLIFFORD, Applications of Grassmann's extensive algebra, *Amer. J. Math.*, **1** (1878), 350–358.

[14] W. K. CLIFFORD, *Mathematical Papers*, ed. R.Tucker (1882; reprinted 1968, Chelsea, New York), pp. 266–276.

[15] A. W. CONWAY, Quaternions and Quantum Mechanics, *Pont. Acad. Sci. Acta,* **12** (1948), 259.

[16] A. W. CONWAY, *Selected Papers*, ed. J. McConnell, Dublin Institute for Advanced Studies, Dublin, 1953, pp. 206–208.

[17] M. J. CROWE, *A History of Vector Analysis: The Evolution of the Idea of a Vectorial System*, University of Notre Dame, Notre Dame, London, 1967.

[18] A. CRUMEYROLLE, *Orthogonal and Symplectic Clifford Algebras: Spinor Structures*, Kluwer Academic Publishers, Dordrecht, Boston, London 1990.

[19] C. DORAN, A. LASENBY, *Geometric Algebra for Physicists*, Cambridge University Press, Cambridge, 2003.

[20] L. DORST, C. DORAN and J. LASENBY, Editors, *Applications of Geometric Algebra in Computer Science and Engineering*, Birkhäuser, Boston, 2002.

[21] P. R. GIRARD, *Quaternions, algèbre de Clifford et physique relativiste*, Presses Polytechniques et Universitaires Romandes, Lausanne, 2004.

[22] P. R. GIRARD, Quaternions, Clifford Algebra and Symmetry Groups, published in *Applications of Geometric Algebra in Computer Science and Engineering*, L. Dorst, C. Doran and J. Lasenby, editors, Birkhäuser, Boston, 2002, pp. 307–315.

[23] P. R. GIRARD, Einstein's equations and Clifford algebra, *Advances in Applied Clifford Algebras*, **9**, 2 (1999), 225–230.

[24] P. R. GIRARD, The quaternion group and modern physics, *Eur. J. Phys.*, **5** (1984), pp. 25–32.

[25] P. R. GIRARD, *The Conceptual Development of Einstein's General Theory of Relativity*, Ph. D. Thesis, University of Wisconsin-Madison, published by Microfilms International, USA, order n° 8105530, 1981.

[26] H. GOLDSTEIN, *Classical Mechanics*, second edition, Addison-Wesley, Reading, Massachusetts, 1980.

[27] K. GÜRLEBECK, W. SPRÖSSIG, *Quaternionic and Clifford Calculus for Physicists and Engineers*, John Wiley & Sons, Chichester, New York, 1997.

[28] F. GÜRSEY and H. C. TZE, Complex and quaternionic analyticity in chiral and gauge theories, *Annals of Physics*, **128** (1980), 29–130.

[29] W. R. HAMILTON, *The Mathematical Papers*, Cambridge University Press, Cambridge, 1931–67, ed. A. W. Conway for the Royal Irish Academy.

[30] W. R. HAMILTON, *Elements of Quaternions*, 2 vols (1899–1901); reprinted Chelsea, New York, 1969.

[31] T. L. HANKINS, *Sir William Rowan Hamilton*, Johns Hopkins University Press, Baltimore and London, 1980.

[32] D. HESTENES, *New Foundations for Classical Mechanics*, 2nd edition, Kluwer Academic Publishers, Dordrecht, Boston, 1999.

[33] D. HESTENES, *Space-Time Algebra*, Gordon and Breach, New York, 1966.

[34] R. D'INVERNO, *Introducing Einstein's Relativity*, Clarendon Press, Oxford, 1992.

[35] B. JANCEWICZ, *Multivectors and Clifford Algebra in Electrodynamics*, World Scientific, Singapore, New Jersey, 1988.

[36] J. B. KUIPERS, *Quaternions and Rotation Sequences: A Primer with Applications to Orbits, Aerospace, and Virtual Reality*, Princeton University Press, Princeton, New Jersey, 1999.

[37] M. LAGALLY, *Vorlesungen über Vektorrechnung*, Akademische Verlagsgesellschaft, Leipzig, 1956, p. 362.

[38] P. LOUNESTO, *Clifford Algebras and Spinors*, second edition, Cambridge University Press, Cambridge, 2001.

[39] R. MEINMNÉ, F. TESTARD, *Introduction à la théorie des groupes classiques*, Hermann, Paris, 1986.

[40] C. MÖLLER, *Relativitätstheorie*, Bibliographisches Institut, B.I.-Wissenschaftsverlag, Wien, Zürich, 1976.

[41] I. R. PORTEOUS, *Clifford Algebras and the Classical Groups*, Cambridge University Press, Cambridge, 1995.

[42] W. RINDLER, *Relativity: Special, General, and Cosmological*, Oxford University Press, Oxford, 2001.

[43] J. B. SHAW, *Vector Calculus with Applications to Physics*, Van Nostrand, New York, 1922.

[44] J. B. SHAW, *Synopsis of Linear Associative Algebra: A Report on its Natural Development and Results up to the Present Time*, Carnegie Institution of Washington, D.C., 1907, pp. 73–74.

[45] L. SILBERSTEIN, *The Theory of Relativity*, Macmillan, London, 1914.

[46] J. SNYGG, *Clifford Algebra, A Computational Tool for Physicists*, Oxford University Press, New York, Oxford, 1995.

[47] F. SOMMEN, Radon and X-ray transforms in Clifford analysis, *Complex variables*, **11** (1989), 49–70.

[48] W. I. STRINGHAM, Determination of the Finite Quaternion Groups, *American Journal of Mathematics*, **4** (1881), 345–357.

[49] J. L. SYNGE, *Quaternions, Lorentz Transformations and the Conway-Dirac-Eddington Matrices*, Dublin Institute for Advanced Studies, Dublin, 1972.

[50] W. THIRRING, *A Course in Mathematical Physics*, Springer-Verlag, New York, Vol. 2, 1992, p. 180.

[51] L. H. THOMAS, *Phil. Mag.*, **3** (1927), 1.

[52] L. H. THOMAS, *Nature*, **117** (1926), 514.

[53] M.-A. TONNELAT, *Les Principes de la théorie électromagnétique et de la relativité*, Masson et Cie, Editeurs, Paris, 1959.

[54] J. VRBIK, Dirac equation and Clifford algebra, *J. Math. Phys.* **35**, 5 (1994), 2309–2314.

Index